AF594987

Photonics for Emerging Applications in Communication and Sensing

Photonics for Emerging Applications in Communication and Sensing

Editors

Guo-Wei Lu
Zhenzhou Cheng
Ting-Hui Xiao

MDPI • Basel • Beijing • Wuhan • Barcelona • Belgrade • Manchester • Tokyo • Cluj • Tianjin

Editors

Guo-Wei Lu
Computer Engineering Division
The University of Aizu
Fukushima
Japan

Zhenzhou Cheng
Department of Optoelectronic
Information Engineering
Tianjin University
Tianjin
China

Ting-Hui Xiao
School of Physics and
Microelectronics
Zhengzhou University
Zhengzhou
China

Editorial Office
MDPI
St. Alban-Anlage 66
4052 Basel, Switzerland

This is a reprint of articles from the Special Issue published online in the open access journal *Photonics* (ISSN 2304-6732) (available at: https://www.mdpi.com/journal/photonics/special_issues/communication_sensing).

For citation purposes, cite each article independently as indicated on the article page online and as indicated below:

LastName, A.A.; LastName, B.B.; LastName, C.C. Article Title. *Journal Name* **Year**, *Volume Number*, Page Range.

ISBN 978-3-0365-8164-4 (Hbk)
ISBN 978-3-0365-8165-1 (PDF)

Contents

About the Editors

Guo-Wei Lu

Guo-Wei Lu obtained his Ph.D. in Information Engineering from the Chinese University of Hong Kong (CUHK) in 2005. He then held a Postdoctoral Fellowship with CUHK from 2005 to 2006. From 2006 to 2009, he worked as an expert researcher at the National Institute of Information and Communications Technology (NICT) in Tokyo, Japan. Afterward, from 2009 to 2010, Dr. Lu served as an assistant professor at Chalmers University of Technology in Gothenburg, Sweden. Subsequently, he worked as a researcher at NICT from 2010 to 2014. From 2014 to 2020, Dr. Lu held the position of associate professor at Tokai University in Japan. Since 2020, he has joined the University of Aizu as a senior associate professor. Additionally, Dr. Lu has been affiliated with NICT as a visiting researcher since 2014 and has served as a visiting lecturer at Kyushu University in Fukuoka, Japan, since 2018. Dr. Lu's research interests include advanced optical modulation formats, electro-optic modulators, photonic signal processing, and optical parametric amplifiers.

Zhenzhou Cheng

Dr. Zhenzhou CHENG (IEEE/Optica Senior Member) is the Chairman and a professor at the Department of Optoelectronic Information Engineering at Tianjin University, as well as an adjunct professor at Georgia Tech Shenzhen Institute. He received his B.Sc. and M.Sc. degrees both from Nankai University. He received his Ph.D. degree in Electronic Engineering from the Chinese University of Hong Kong under the support of the Hong Kong Ph.D. Fellowship Scheme. After two years of postdoctoral research, he joined Goda Lab in the Department of Chemistry at the University of Tokyo as an assistant professor. In 2018, he joined the School of Precision Instruments and Optoelectronics Engineering at Tianjin University as a full professor. His research interests focus on silicon photonics. He is in charge of several projects funded by NSFC and JSPS and has published over 100 peer-reviewed SCI papers in top-tier academic journals, namely, *Nature Photonics*.

Ting-Hui Xiao

Dr. Tinghui Xiao is a professor at the Department of Physics and the Director at the Institute of Optoelectronic Information Science at Zhengzhou University. He received his B.Sc. degree from Sun Yat-sen University in 2013, M.Sc. degree from Institute of Physics, Chinese Academy of Sciences in 2016, and Ph.D. degree from the University of Tokyo in 2019. He worked at the University of Tokyo as an assistant professor from 2019 to 2022 and then joined Zhengzhou University as a professor. His research interests focus on silicon photonics, vibrational spectroscopy, and ultrafast imaging.

Editorial
Special Issue "Photonics for Emerging Applications in Communication and Sensing"

Guo-Wei Lu [1,*], Zhenzhou Cheng [2] and Ting-Hui Xiao [3,4]

1 Division of Computer Engineering, The University of Aizu, Fukushima 965-8580, Japan
2 School of Precision Instruments and Optoelectronics Engineering, Tianjin University, Tianjin 300072, China; zhenzhoucheng@tju.edu.cn
3 Henan Key Laboratory of Diamond Optoelectronic Materials and Devices, School of Physics and Microelectronics, Zhengzhou University, Zhengzhou 450052, China; xiaoth@zzu.edu.cn
4 Institute of Quantum Materials and Physics, Henan Academy of Sciences, Zhengzhou 450046, China
* Correspondence: gordon.guoweilu@gmail.com

Citation: Lu, G.-W.; Cheng, Z.; Xiao, T.-H. Special Issue "Photonics for Emerging Applications in Communication and Sensing". *Photonics* **2023**, *10*, 738. https://doi.org/10.3390/photonics10070738

Received: 19 June 2023
Accepted: 26 June 2023
Published: 28 June 2023

Photonics has emerged as a crucial enabler for various emerging applications in communication and sensing, revolutionizing industries such as data centers, autonomous driving, 5G wireless networks, cloud computing, the IoT, and virtual reality. This Special Issue aims to present recent advancements and address future challenges in photonics technologies, focusing on performance monitoring in optical networks, photonic sensors and devices, optical signal processing subsystems, and digital signal processing for future optical transmission systems. This collection of research consists of thirteen articles highlighting the transformative potential of photonics in supporting and enabling the advancement of emerging technologies in communication and sensing. By exploring these frontiers, we strive to uncover innovative solutions that will shape the future of communication and sensing.

In optical networks, it is essential to estimate the quality of transmission (QoT) of light paths and monitor the performance of each wavelength channel. Y. Zhou et al. presented a self-attention mechanism-based multi-channel QoT estimator for optical networks. Based on the previous QoT estimation scheme using an artificial neural network (ANN), a self-attention mechanism was introduced to dynamically assign weights to the interfering channels to improve the estimation accuracy and scalability. Specifically, the maximal absolute error achieved by SA-QoT-E was reduced from 4.21 to 1.13 dB in the Japan network (Jnet) and from 3.46 to 0.84 dB in the National Science Foundation network (NSFnet) [1]. To efficiently monitor the optical performance in wavelength division (WDM) networks with low implementation costs, T. Yang et al. introduced optical labels encoded in differential phase-shift keying (DPSK) to enable simultaneous monitoring of all wavelength channels using a single low-bandwidth photodetector (PD) and designed digital signal processing (DSP) algorithms. Less than 1 dB error was achieved after 20-span WDM transmission, verifying the feasibility and efficiency of the proposed scheme [2]. To enhance the physical layer security of optical networks, the signal-to-noise ratio (SNR) of the optical channel can be extracted and used as the basis of key generation. X. Wang et al. proposed a cost-effective key distribution scheme using the extracted SNR as a key for secured communication with promising metrics such as a key generation rate of 25 kbps and a key consistency rate of 98% [3].

Plasmonic sensors are photonics-based devices that exploit the interaction between light and plasmons and the collective oscillations of electrons to detect and analyze changes in local electromagnetic fields. J. Liu et al. investigated the optical extinction spectra of a silver nanocube driven by an ultrashort carrier-envelope phase (CEP)-locked laser pulse. They revealed five localized surface plasmon resonance (LSPR) modes and their physical origin, demonstrating the modulation capability of CEP in selectively exciting LSPR modes. This research provides insights into controlling and optimizing the optical properties of metal nanoparticles for sensing and detection applications [4]. While traditionally associated with noble metals like gold and silver, plasmonic sensors can also be implemented by using alternative materials such as aluminum nitride (AlN). S. Tang et al. explored a dielectric-grating-assisted plasmonic sensor utilizing the bound states in the continuum (BIC) effect for narrow line widths and high sensitivity in refractive index sensing. This approach addresses the limitations of plasmonic sensors and opens possibilities for improved sensing technologies [5].

Emerging materials such as silicon-based compounds, metasurfaces, and phase-change materials have enabled exciting advancements in photonic devices for optical signal processing. Y. Hong et al. presented an optimized approach using the 3D finite element method to achieve flat and low dispersion profiles in silicon nitride waveguides for efficient four-wave mixing (FWM). Their work results in near-zero dispersion values and high conversion efficiencies with low pump power, enhancing the performance of optical signal processing applications based on FWM [6]. X. Tian et al. introduced a hybrid phase-change metasurface carpet cloak using a novel phase-change material (GSST). This metasurface offers wideband indiscernibility, polarization-immune cloaking, and reversible functionality, providing potential applications in electromagnetic camouflage and illusion [7].

Optical signal processing subsystems play a pivotal role in optical communication, focusing on developing innovative techniques to manipulate, analyze, and enhance optical signals, enabling efficient transmission, detection, and processing of information in optical transmission systems. Y. Liu et al. proposed an optical analog-to-digital converter (OADC) scheme using a multimode interference (MMI) coupler for enhanced bit resolution. Their approach achieved 20 quantization levels with only 6 channels, demonstrating robustness and integration potential [8]. A. Chiba et al. achieved 60 GHz separation optical two-tone signal generation at arbitrary C-band wavelengths without complicated optical wavelength filtering. Their method selectively suppresses undesired low-order optical sidebands using a polarizer, resulting in more than 20 dB suppression over a 40 nm wavelength range [9]. Z. Liu et al. demonstrated a high-resolution 1×6 programmable photonics spectral processor (PSP) using low-cost, compact spatial light paths. This processor allows independent manipulation of filtering bandwidth and power attenuation for each wavelength channel, offering potential benefits for WDM systems [10].

DSP plays a crucial role in extracting, decoding, and enhancing optical signals, ensuring reliable transmission, coherent detection, and efficient utilization of the available optical bandwidth in both coherent and intensity-modulation direct-detection (IMDD) optical communication systems. It encompasses a wide range of techniques and algorithms that enable the optimization of signal quality, advanced modulation formats, error correction, and compensation of impairments. T. Yang et al. proposed a low-complexity scheme for accurate chromatic dispersion (CD) estimation in faster-than-Nyquist (FTN) coherent optical transmission systems. Their scheme achieves high CD estimation accuracy (65 ps/nm) with reduced computational complexity (3%) compared to conventional methods, making it suitable for practical FTN systems [11]. Y. Zhang et al. proposed a parallel distribution matcher scheme for probabilistic-shaped coherent optical fiber communication. The SNR can be enhanced by 0.12 dB, and the block length can be reduced by 40% compared to the conventional scheme using a constant composition distribution matcher (CCDM) [12]. Y. Liu et al. investigated signal pre-processing schemes for SNR equalization in short-reach probabilistically shaped discrete multitone (DMT) transmission systems. The proposed fast Walsh–Hadamard transform (WHT)-based pre-distortion scheme offers a receiver sensitivity gain of 1 dB [13].

This Special Issue on photonics technologies showcases remarkable advancements and prospects in the field. The featured research articles highlight innovative solutions and address key challenges in various topics, such as optical performance monitoring, photonic sensors, photonic devices for optical signal processing, optical signal processing subsystems, and digital signal processing. The continued development of photonics for emerging applications in communication and sensing holds great promise for the future.

Conflicts of Interest: The authors declare no conflict of interest.

References

1. Zhou, Y.; Huo, X.; Gu, Z.; Zhang, J.; Ding, Y.; Gu, R.; Ji, Y. Self-Attention Mechanism-Based Multi-Channel QoT Estimation in Optical Networks. *Photonics* **2023**, *10*, 63. [CrossRef]
2. Yang, T.; Li, K.; Liu, Z.; Wang, X.; Shi, S.; Wang, L.; Chen, X. Optical Labels Enabled Optical Performance Monitoring in WDM Systems. *Photonics* **2022**, *9*, 647. [CrossRef]
3. Wang, X.; Zhang, J.; Wang, B.; Zhu, K.; Song, H.; Li, R.; Zhang, F. Key Distribution Scheme for Optical Fiber Channel Based on SNR Feature Measurement. *Photonics* **2021**, *8*, 208. [CrossRef]
4. Liu, J.; Li, Z. Control of Surface Plasmon Resonance in Silver Nanocubes by CEP-Locked Laser Pulse. *Photonics* **2022**, *9*, 53. [CrossRef]
5. Tang, S.; Chang, C.; Zhou, P.; Zou, Y. Numerical Study on a Bound State in the Continuum Assisted Plasmonic Refractive Index Sensor. *Photonics* **2022**, *9*, 224. [CrossRef]
6. Hong, Y.; Hong, Y.; Hong, J.; Lu, G.-W. Dispersion Optimization of Silicon Nitride Waveguides for Efficient Four-Wave Mixing. *Photonics* **2021**, *8*, 161. [CrossRef]
7. Tian, X.; Xu, J.; Xiao, T.-H.; Ding, P.; Xu, K.; Du, Y.; Li, Z.-Y. Broadband Generation of Polarization-Immune Cloaking via a Hybrid Phase-Change Metasurface. *Photonics* **2022**, *9*, 156. [CrossRef]
8. Liu, Y.; Qiu, J.; Liu, C.; He, Y.; Tao, R.; Wu, J. An Optical Analog-to-Digital Converter with Enhanced ENOB Based on MMI-Based Phase-Shift Quantization. *Photonics* **2021**, *8*, 52. [CrossRef]
9. Chiba, A.; Akamatsu, Y. Wavelength-Tunable Optical Two-Tone Signals Generated Using Single Mach-Zehnder Optical Modulator in Single Polarization-Mode Sagnac Interferometer. *Photonics* **2022**, *9*, 194. [CrossRef]
10. Liu, Z.; Li, C.; Tao, J.; Yu, S. Programmable High-Resolution Spectral Processor in C-Band Enabled by Low-Cost Compact Light Paths. *Photonics* **2020**, *7*, 127. [CrossRef]
11. Yang, T.; Jiang, Y.; Wang, Y.; You, J.; Wang, L.; Chen, X. Low-Complexity and Highly-Robust Chromatic Dispersion Estimation for Faster-than-Nyquist Coherent Optical Systems. *Photonics* **2022**, *9*, 657. [CrossRef]
12. Zhang, Y.; Wang, H.; Ji, Y.; Zhang, Y. Parallel Distribution Matcher Base on CCDM for Probabilistic Amplitude Shaping in Coherent Optical Fiber Communication. *Photonics* **2022**, *9*, 604. [CrossRef]
13. Liu, Y.; Long, H.; Chen, M.; Cheng, Y.; Zhou, T. Probabilistically-Shaped DMT for IM-DD Systems with Low-Complexity Fast WHT-Based PDSP. *Photonics* **2022**, *9*, 655. [CrossRef]

 *photonics*

Article

Self-Attention Mechanism-Based Multi-Channel QoT Estimation in Optical Networks

Yuhang Zhou [1], Xiaoli Huo [2], Zhiqun Gu [1,*], Jiawei Zhang [1], Yi Ding [2], Rentao Gu [1] and Yuefeng Ji [1]

1 State Key Lab of Information Photonics and Optical Communications, Beijing University of Posts and Telecommunications (BUPT), Beijing 100876, China
2 China Telecom Research Institute, Beijing 102209, China
* Correspondence: guzhiqun@bupt.edu.cn

Abstract: It is essential to estimate the quality of transmission (QoT) of lightpaths before their establishment for efficient planning and operation of optical networks. Due to the nonlinear effect of fibers, the deployed lightpaths influence the QoT of each other; thus, multi-channel QoT estimation is necessary, which provides complete QoT information for network optimization. Moreover, the different interfering channels have different effects on the channel under test. However, the existing artificial-neural-network-based multi-channel QoT estimators (ANN-QoT-E) neglect the different effects of the interfering channels in their input layer, which affects their estimation accuracy severely. In this paper, we propose a self-attention mechanism-based multi-channel QoT estimator (SA-QoT-E) to improve the estimation accuracy of the ANN-QoT-E. In the SA-QoT-E, the input features are designed as a sequence of feature vectors of channels that route the same path, and the self-attention mechanism dynamically assigns weights to the feature vectors of interfering channels according to their effects on the channel under test. Moreover, a hyperparameter search method is used to optimize the SA-QoT-E. The simulation results show that, compared with the ANN-QoT-E, our proposed SA-QoT-E achieves higher estimation accuracy, and can be directly applied to the network wavelength expansion scenarios without retraining.

Keywords: quality of transmission (QoT) estimation; nonlinear effect; multi-channel; self-attention mechanism

Citation: Zhou, Y.; Huo, X.; Gu, Z.; Zhang, J.; Ding, Y.; Gu, R.; Ji, Y. Self-Attention Mechanism-Based Multi-Channel QoT Estimation in Optical Networks. *Photonics* **2023**, *10*, 63. https://doi.org/10.3390/photonics10010063

Received: 31 October 2022
Revised: 27 November 2022
Accepted: 3 January 2023
Published: 6 January 2023

1. Introduction

Precise and fast estimation of quality of transmission (QoT) for a lightpath prior to its deployment becomes capital for network optimization [1]. Traditional physical layer model (PLM)-based QoT estimation utilizes optical signal transmission theories to predict the lightpaths' QoT values, which mainly includes two approaches: (1) sophisticated analytical models, such as the split-step Fourier method [2]; (2) approximated analytical models, such as the Gaussian Noise (GN) model [3]. The former approach analyzes various physical layer impairments and provides high estimation accuracy, but it requires high computational resources. Thus, sophisticated analytical models are unable to estimate the lightpaths' QoT values online and are not scalable to large-scale networks and dynamic network scenarios [2], and the latter approach, of which the widely used GN model simplifies the nonlinear interference (NLI) to additive Gaussian noise, provides quick QoT estimation for lightpaths, but it requires an extra design margin to ensure the reliability of the lightpaths' transmission in the worst-case, thus leading to underutilization of network resources [3]. The PLM-based QoT estimation cannot ensure high accuracy and low computational complexity simultaneously.

Machine learning (ML) has powerful data mining and fast prediction abilities, it has been widely used for QoT estimation. Most of the existing studies focus on the QoT estimation for a signal lightpath [4–20], and all of these achieve high accuracy. For example, in [4], three ML-based classifiers, which include random forest (RF), support vector machine

(SVM), and K-nearest neighbor (KNN), are proposed to predict the QoT of lightpaths. The simulation results show that these classifiers achieve high accuracy and the SVM-based classifier achieves the best performance with a classification accuracy of 99.15%. In [9], a neural network (NN)-based QoT regressor is proposed, and the experimental results show that the regressor achieves a 90% optical signal-to-noise (OSNR) prediction of a 0.25 dB root-mean-squared-error (RMSE) on a mesh network. In [15], a meta-learning assisted training framework for an ML-based PLM is proposed, and the framework can improve the model robustness to uncertainties of the parameters and enable the model to converge with few data samples. In [20], it is explored how ML can be used to achieve lightpath QoT estimation and forecast tasks, and the data processing strategies are discussed with the aim to determine the input features of the QoT classifiers and predictors.

However, due to the nonlinear effect of fibers, the newly deployed lightpaths (new-LPs) degrade the QoT of previously deployed lightpaths (pre-LPs). The single-channel (lightpath) QoT estimation only provides the QoT information of the new-LPs, which leads to the decrease in QoT estimation accuracy of the pre-LPs. Thus, it is necessary to estimate the QoT of new-LPs and pre-LPs simultaneously, i.e., multi-channel QoT estimation. In [21], a deep grapy convolutional neural network (DGCNN)-based QoT classifier is proposed, of which the aim is to accurately classify any unseen lightpath state. In [22], a novel deep convolutional neural network (DCNN)-based QoT classifier is proposed for network-wide QoT estimation. The above works achieve high accuracy of multi-channel QoT classification. Nevertheless, the QoT classifier cannot provide detailed QoT values of lightpaths; thus, it cannot be directly applied to network planning, such as modulation format assignment. Reference [23] proposes an ANN-based multi-channel QoT estimator (ANN-QoT-E) over a 563.4 km field-trial testbed, and the mean absolute error (MAE) is about 0.05 dB for the testing data. However, in optical networks, there exist three NLIs, i.e., self-channel interference (SCI), cross-channel interference (XCI), and multi-channel interference (MCI). When the interfering channel with a higher power is closer to the spectrum of the channel under test (CUT), stronger NLI is introduced on the CUT [3]. Thus, the effects of the interfering channels on the CUTs' QoT are different. The ANN-QoT-E proposed in [23] neglects the different effects of the interfering channels in its input layer, which affects its accuracy. Therefore, how to assign different weights to interfering channels for the QoT estimation of CUTs to enhance the accuracy becomes a crucial problem.

In this paper, we extend our previous work in [24], which applies self-attention mechanisms to improve the accuracy of the ANN-QoT-E. The comparison of this work with the previous works is shown in Table 1, and the main contributions of this paper are summarized as follows:

(1) We propose a self-attention mechanism-based multi-channel QoT estimator (SA-QoT-E) to improve the estimation accuracy of the ANN-QoT-E, where the input features are designed as a sequence of channel feature vectors, and the self-mechanism dynamically assigns weights to the interfering channels for the QoT estimation of the CUTs.

(2) We use a hyperparameter search method to optimize the SA-QoT-E, which selects optimal hyperparameters for the SA-QoT-E to improve its estimation accuracy.

(3) We show the performance of the SA-QoT-E via extensive simulations. The simulation results show that the SA-QoT-E achieves a higher estimation accuracy than the ANN-QoT-E proposed in [23], and it is verified that the assignment of attention weights to the interfering channels conforms to the optical transmission theory. Compared with the ANN-QoT-E, the SA-QoT-E is more scalable, and it can be directly applied to network wavelength expansion scenarios without retraining. By analyzing the computational complexity of the SA-QoT-E and the ANN-QoT-E, it is concluded that the SA-QoT-E has higher computational complexity. However, the training phase of QoT estimators is offline and the computational complexity of the SA-QoT-E is acceptable; thus, the SA-QoT-E still has more advantages than the ANN-QoT-E in practical network applications.

Table 1. Comparison of this work with the previous works.

Work	QoT Classification	QoT Regression	Channel Effect Quantification	Hyperparameters Search
Ref [21]	✓			
Ref [22]	✓			
Ref [23]		✓		
Ref [24]		✓	✓	
This work		✓	✓	✓

The remainder of this paper is organized as follows. In Section 2, we describe the principle of the self-attention mechanism-based multi-channel QoT estimation scheme. The dataset generation and simulation setup are shown in Section 3. In Section 4, we discuss the simulation results in terms of convergence, estimation accuracy, scalability, and computational complexity. Finally, we make a conclusion in Section 5.

2. Self-Attention Mechanism-Based Multi-Channel QoT Estimation

The self-attention mechanism improves the accuracy of the task model by calculating the relevance between Query (Q: query state) and Key (K: candidate state), allowing the task model to dynamically pay attention to the input feature vectors and extract more important information [25]. To solve the problem that ANN models ignore the different effects of the interfering channels in their inputs, we apply the self-attention mechanism to assign the dynamic weights to the input channel features, i.e., SA-QoT-E.

The schematic diagram of the SA-QoT-E is shown in Figure 1a. The steps of the SA-QoT-E are shown as follows:

(1) Network database collection: The network database, which includes the transmission configuration (such as the launch power of each channel, the wavelength allocation, and transmission distance) and the corresponding QoT value of each channel, is collected from the network topology first.
(2) Channel features input: Then, the channel features from the network database are input to the self-attention mechanism block.
(3) New channel features generation: The self-attention mechanism assigns weights to the input channel features according to their effects on the CUT. Thus, the new channel features with interfering channel information (ICI) are generated.
(4) Channels QoT estimation: Finally, the channel features with ICI are input into an ANN model to estimate the QoT values of the channels.

In the SA-QoT-E, the database generation in step (1) is shown in Section 3 and the ANN model in step (4) is a classic ML model [26]. Thus, in this section, we introduce the design of the input of the self-attention mechanism in step (2) and the process of the self-attention mechanism in step (3).

2.1. The Design of the Input of Self-Attention Mechanism

The self-attention mechanism takes a sequence of feature vectors as input. For a certain feature vector in the sequence, the self-attention mechanism dynamically assigns a weight to each feature vector according to its effect on the vector under test, and finally generates a new feature vector containing the entire sequence information. In optical transmission systems, due to the nonlinear effect of fibers, the characteristics of the interfering channels (such as the launch power and the wavelength) decide their effects on the QoT of the CUT, where the interfering channels and the CUT pass through the same route. Therefore, in the SA-QoT-E, we design the input of the self-attention mechanism as a sequence of channel feature vectors, where the channels pass through the same route and the channel feature vectors contain the corresponding channel transmission characteristics. Each channel feature vector in a sequence is shown as follows:

$$F_i = [p_i, w_i, l_i], \tag{1}$$

where F_i represents the channel feature vector of the i-th channel, p_i is the launch power of the i-th channel, w_i is the wavelength of the i-th channel, and l_i is the transmission distance of the i-th channel.

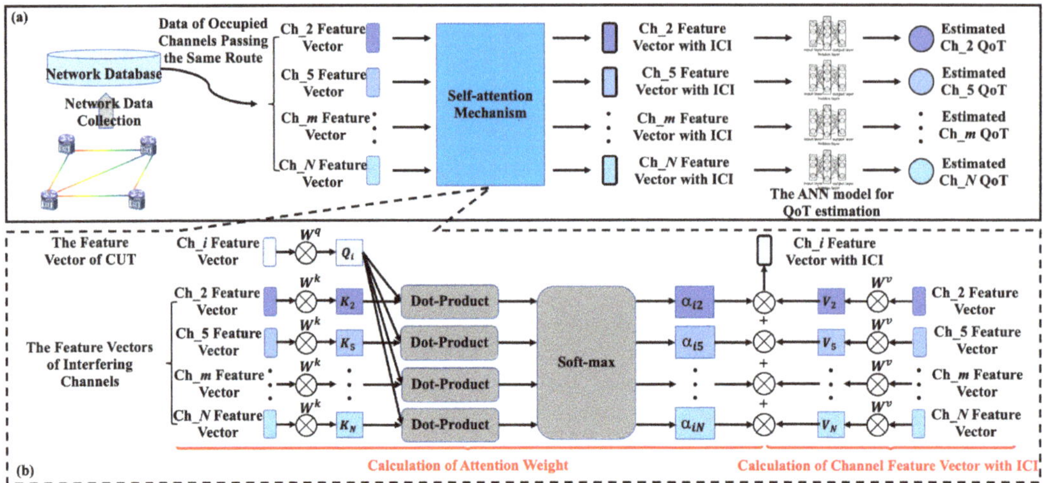

Figure 1. (**a**) Schematic diagram of self-attention mechanism-based multi-channel QoT estimation scheme; (**b**) process of self-attention mechanism.

2.2. The Process of Self-Attention Mechanism Block

In the self-attention mechanism of the SA-QoT-E, the attention weights represent the effects of the interfering channels. The higher the attention weight of the interfering channel, the more the SA-QoT-E pays attention to the corresponding interfering channel. There are three parameter matrices trained from the training data, which are W^q, W^k, and W^v, and the channel feature vectors are multiplied by these matrices to obtain the corresponding Q, K, and V. Then, the attention weight is a function of Q and K, and the new channel feature vector with ICI is calculated by the weighted summation of the attention weight and V. The self-attention mechanism block is shown in detail in Figure 1b. There are two main parts in the block: (1) the calculation of attention weight; (2) the calculation of channel feature vector with ICI.

2.2.1. The Calculation of Attention Weight

As shown in Figure 1b, the feature vectors of the i-th channel, which is the CUT, and the interfering channels, which also include the CUT itself, are used for the calculation of Q and K, respectively. For the calculation of attention weight, the two most commonly used functions are Dot-product and Additive [25]. The two functions have similar computational complexity; however, Dot-product can be implemented using highly optimized matrix multiplication, and it is much faster and more space-efficient in practice than Additive. In this work, Dot-product is chosen as the function to calculate the attention weight, and the soft-max function is applied for the normalization of attention weights. The attention weight α_{ij} can be calculated as follows:

$$\alpha_{ij} = softmax\left(F_i W^q \left(F_j W^k\right)^T\right), \tag{2}$$

where α_{ij} is the attention weight of interfering channel j for the CUT i; $softmax(\cdot)$ is the soft-max function; W^q and W^k are the $n_d \times n_d$ (n_d is the dimension of the channel feature vectors, $n_d = 3$ in this paper) trainable parameter matrices for the calculation of Q and K, respectively; F_i and F_j are the feature vectors of channel i and channel j, respectively.

When the attention weights of the interfering channels for the CUT i are obtained by Formula (2), the new feature vector with ICI of channel i can be calculated for the QoT estimation of channel i.

2.2.2. The Calculation of Channel Feature Vector with ICI

The attention weights of the interfering channels are the effects of the interfering channels on the CUT, and the channel feature vector with ICI is calculated as follows:

$$F_i' = \sum_j \alpha_{ij} F_j W^v, \tag{3}$$

where F_i' is the feature vector with ICI of channel i, α_{ij} is the attention weight of interfering channel j for the CUT i, F_j is the feature vector of the interfering channel j, and W^v is the $n_d \times n_d$ trainable parameter matrix for the calculation of V.

The channel feature vector with ICI F_i' contains the information of CUT i and all the interfering channels, and the channel feature vector F_i' is input to an ANN model to estimate the QoT value of the CUT i. Similarly, we can obtain the QoT values of all channels to achieve the multi-channel QoT estimation task.

3. Data Generation and Simulation Setup

In the SA-QoT-E, the input is the feature vectors of all occupied channels passing the same route, and the output is the generalized signal-to-noise ratio (GSNR) values of these channels. The GSNR can be calculated as follows:

$$GSNR = \frac{P}{P_{ASE} + P_{NLI}}, \tag{4}$$

where P is the launch power of the lightpath, P_{ASE} is the amplified spontaneous emission (ASE) noise power introduced by erbium-doped fiber amplifiers (EDFAs), and P_{NLI} is the NLI power due to the nonlinear effect of fibers. The Japan network (Jnet) and National Science Foundation network (NSFnet) are considered in our simulations, which are shown in Figure 2a,b, respectively. The transponders of the two networks are set to work on the C++ band of which the center frequency is 193.35 THz with a spectral load with 80 wavelengths (i.e., channels) on the 50 GHz spectral grid. The symbol rate of the transponders is 32 GBaud, and the launch power of each channel is uniformly selected in the range of [−3~0] dBm with 0.1 dBm granularity. We assume the fibers in the two networks are ITU-T G.652 standard single-mode fibers (SSMF), of which attenuation, dispersion, and non-linearity coefficients are 0.2 dB/km, 16.7 ps/nm/km, and 1.3/W/km, respectively; the span length of the fibers is 40 km in the Jnet and 100 km in the NSFnet. The EDFas in the two networks are set to completely compensate for fiber span losses, and the noise figure is 8 dB in the Jnet and 6.5 dB in the NSFnet. The network datasets of the two networks are generated synthetically by the open-source Gaussian noise model in the Python (GNPY) library [27], where the NLI power is calculated by the generalized Gaussian model (GNN). In each network, we generate 8000 samples for training and 2000 samples for testing by randomly choosing one of the K shortest paths of a source–destination node pair and a channel state that represents whether the channels are occupied. The training datasets in the Jnet and NSFnet are defined as D^{80}_{Jnet} and D^{80}_{NSFnet}, and the testing datasets in the Jnet and NSFnet are defined as T^{80}_{Jnet} and T^{80}_{NSFnet}. Each sample of the network dataset contains the launch power of each channel, the wavelength allocation, the transmission length, and the GSNR of each channel.

In our simulations, the performance comparison of the ANN-QoT-E and the SA-QoT-E is shown. The input of the ANN-QoT-E contains the launch power of each channel (80-dimensional vector), the wavelength allocation (80-dimensional vector), and the transmission distance, and the outputs of the ANN-QoT-E are the GSNR values of all channels (80-dimensional vector). Thus, the sizes of the first and the last layers of the ANN-QoT-E are 161 and 80, respectively. In the ANN model of the SA-QoT-E, its input is the channel feature vector with ICI (3-dimensional vector) and its output is the GSNR value of the channel. Thus, the sizes of the first and the last layers of the ANN model of the SA-QoT-E are 3 and 1, respectively.

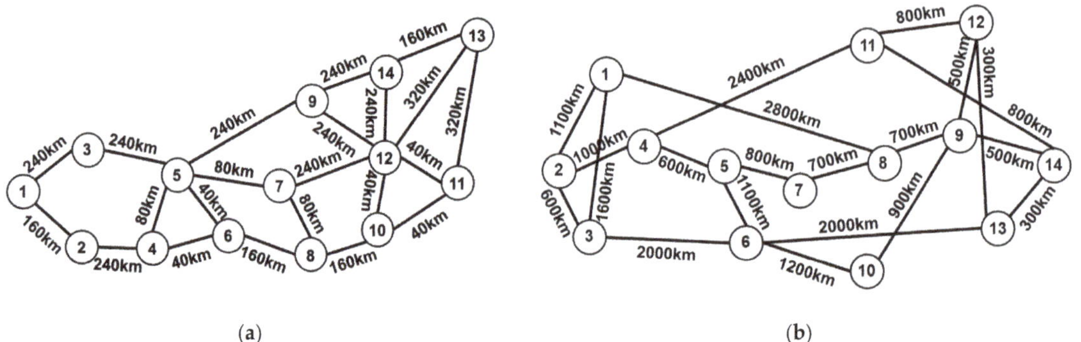

Figure 2. (**a**) Japan network (Jnet) topology; (**b**) National Science Foundation network (NSFnet) topology.

We optimize the hyperparameters of the SA-QoT-E to achieve high accuracy. To reduce the operation time of the hyperparameters search method, we empirically select the set of available hyperparameters rather than all of those. In the SA-QoT-E, we set the number of heads of the self-attention mechanism N_s, the number of hidden layers N_L, the number of neurons in hidden layers N_h, the batch size b, the learning rate α, and the epoch number e as variables with the constraints that $N_s \in \{1, 2, 3, 4\}$, $N_L \in \{1, 2, 3, 4\}$, $N_h \in \{32, 64, 128, 256, 512\}$, $b \in \{16, 32, 64, 128\}$, $\alpha \in \{0.1, 0.01, 0.001\}$, and $e \in \{100, 200, 400, 600, 800\}$. The hyperparameters of the ANN-QoT-E are similar to the ANN model of the SA-QoT-E.

After searching for the hyperparameters to achieve the highest accuracy on training datasets, the number of heads of the self-attention mechanism is set to 1 in the SA-QoT-E. The ANN-QoT-E and the ANN model of the SA-QoT-E are set to be composed of fully connected layers including 161/256/256/80 and 3/256/256/1 neurons, respectively. The activation function for all neurons is the rectified linear unit (ReLU) function and the loss function is the mean square error (MSE). In the training phase, the batch size, the learning rate, and the number of epochs of the ANN-QoT-E and the SA-QoT-E are 32, 0.01, and 400, respectively. We extract 1/10 of the training set as the verification set, and the trained mode is verified every 50 epochs and the current optimal model is saved.

4. Simulation Results and Analyses

Figure 3a,b show the GSNR decrease of the Ch_1, Ch_20, Ch_40, Ch_60, and Ch_80 caused by the deployment of the new lightpaths in the Jnet and the NSFnet. In the Jnet and the NSFnet, the selected routes are 1-3-5-9-14-13 and 2-4-11-12, respectively, and the wavelength assignment scheme is the first-fit (FF). In Figure 3a, with the increase in the number of newly deployed lightpaths, the GSNR decrease of the CUTs increases and the maximum GSNR decrease achieves 3.25 dB. Moreover, due to the FF scheme, the GSNR of the channel with the smallest number (Ch_1) decreases the most when the number of newly deployed lightpaths is 10, and when the number of newly deployed lightpaths achieves 70, the GSNR of the channel with the middle number (Ch_40) decreases the most. In Figure 3b, the performance of the GSNR decrease in the NSFnet is similar to that in the Jnet, and the maximal GSNR decrease of the CUTs is 1.36 dB. The QoT of the previously deployed lightpaths is deteriorated greatly due to the deployment of new lightpaths, and the QoT estimation of the single lightpath cannot capture the decrease in the previously deployed lightpaths' QoT. Thus, it is necessary to estimate the QoT of new-LPs and pre-LPs simultaneously, i.e., achieve multi-channel QoT estimation, in optical networks.

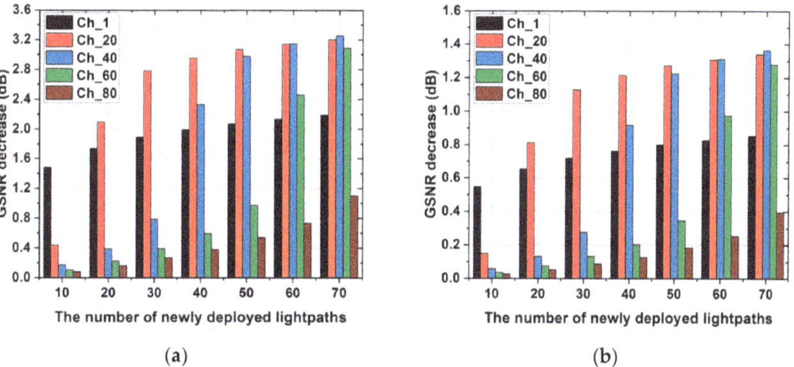

Figure 3. (**a**) GSNR decrease in the Jnet; (**b**) GSNR decrease in the NSFnet.

In this section, we show the performance of the ANN-QoT-E proposed in [23] and our proposed SA-QoT-E in respect of model convergence, estimation accuracy, scalability, and computational complexity.

4.1. Convergence

The training processes of the ANN-QoT-E and the SA-QoT-E in the Jnet and NSFnet are shown in Figure 4a,b. As shown in Figure 4a, in the Jnet, the ANN-QoT-E and the SA-QoT-E converge after 400 epochs. After 79 epochs, the ANN-QoT-E almost converges and its training loss is 0.127, and the SA-QoT-E almost converges at the 51th epoch and its training loss is 0.127. In Figure 4b, after 400 epochs, the two models converge in the NSFnet, where the SA-QoT-E almost converges at the 53th epoch with the training loss of 0.059 and the ANN-QoT-E almost converges at 226th epoch with the training loss of 0.057. In conclusion, the SA-QoT-E converges in fewer epochs than the ANN-QoT-E. That is because the SA-QoT-E has a stronger data-fitting ability than the ANN-QoT-E.

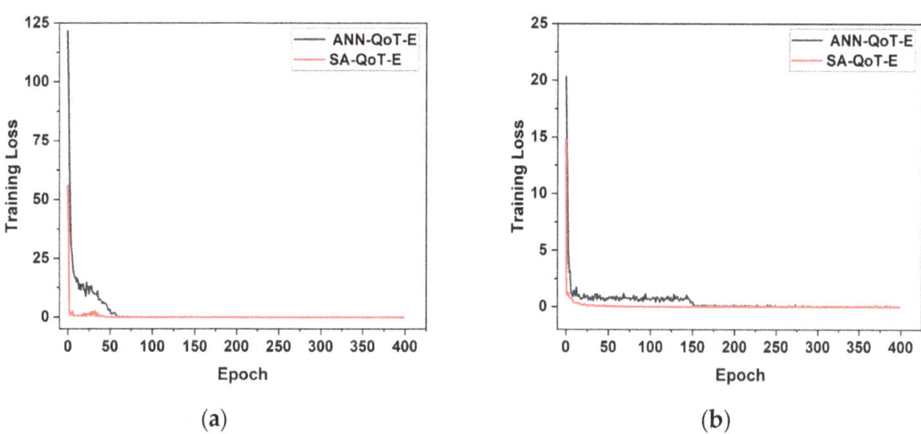

Figure 4. (**a**) Training loss vs. the number of epochs for the ANN-QoT-E and the SA-QoT-E in the Jnet; (**b**) training loss vs. the number of epochs for the ANN-QoT-E and the SA-QoT-E in the NSFnet.

4.2. Estimation Accuracy

Figure 5 shows the estimation accuracy of the ANN-QoT-E and the SA-QoT-E tested on the testing dataset. Figure 5a shows the testing of the mean absolute error (MAE) of the two estimation models in the Jnet and NSFnet. The testing MAE of the SA-QoT-E is lower than that of the ANN-QoT-E; specifically, the testing MAE of the SA-QoT-E is 0.147 dB and 0.127 dB lower than that of the ANN-QoT-E in the Jnet and NSFnet, respectively, which means the predicted GSNR of the SA-QoT-E is, on average, 0.147 dB and 0.127 dB closer to the ground truth of the GSNR compared with that of the ANN-QoT-E in the Jnet and NSFnet, respectively. R2 score is a commonly used metric for the evaluation of the regression model, it is closer to 1, and the corresponding model has higher estimation accuracy. Figure 5b shows that, in the Jnet and the NSFnet, the R2 score of the SA-QoT-E is 0.03 and 0.016 higher than that of the ANN-QoT-E, respectively.

Figure 5c,d show the predicted GSNR of the ANN-QoT-E and the SA-QoT-E against their actual GSNR in the Jnet, respectively. The ideal estimation result is that the scatters distribute on the baseline, which means the predicted GSNR is equal to the actual GSNR. The figures show that, in the Jnet, the estimation accuracy of the SA-QoT-E is higher than that of the ANN-QoT-E; specifically, the maximum absolute error of the SA-QoT-E is 1.13 dB and that of the ANN-QoT-E is 4.21 dB.

Similarly, Figure 5e,f show the predicted GSNR values of the two models against their actual GSNR in the NSFnet. The results show that the estimation accuracy of the SA-QoT-E is higher than that of the ANN-QoT-E in the NSFnet, where the maximum absolute error of the SA-QoT-E is 0.84 dB and that of the ANN-QoT-E is 3.46 dB.

In the Jnet and the NSFnet, the accuracy of the SA-QoT-E is higher than that of the ANN-QoT-E, which is because the self-attention mechanism assigns weights to the interfering channels according to their effects on the QoT of CUT. Figure 6a,b show the attention weight and launch power of each channel for Ch_16 in the Jnet and for Ch_76 in the NSFnet, respectively, which is tested by the SA-QoT-E on a randomly selected testing sample. Due to the nonlinear effect of fibers, the interfering channel, which has a higher launch power and closer spectral distance to the CUT (SDC), has a stronger effect on the CUT. Figure 6a marks the channels with the larger attention weight. In the Jnet, as shown in Figure 6a, these marked channels are arranged in descending order of attention weight as Ch_16, Ch_7, Ch_14, Ch_75, Ch_72, Ch_13, Ch_60, and Ch_25. The attention weight of Ch_16 is maximal due to its launch power being maximal and SDC being minimal. Though the SDC of Ch_7 is higher than that of Ch_14, the attention weight of Ch_7 is higher than that of Ch_14 due to the higher launch power of Ch_7. The attention weights assigned for Ch_13, Ch_25, Ch_60, Ch_72, and Ch_75 violate the nonlinear effect theory; thus, the accuracy of the SA-QoT-E in the Jnet is lower than that in the NSFnet. Figure 6b shows that, in the NSFnet, the order of the marked channels according to their attention weight is Ch_76, Ch_29, Ch_10, Ch_69, and Ch_41. The attention weight of Ch_76 is maximal due to the maximal launch power and minimal SDC of Ch_76, and the attention weight of Ch_29 and Ch_10 is higher than that of Ch_69 and Ch_41, which is because the launch power of Ch_29 and Ch_10 is higher than that of Ch_69 and Ch_41. Due to the smaller SDC of Ch_29 than Ch_10, the attention weight of Ch_29 is higher than that of Ch_10. For the same reason, the attention weight of Ch_69 is higher than that of Ch_41. The attention weights assignment in the NSFnet obeys the nonlinear effect theory.

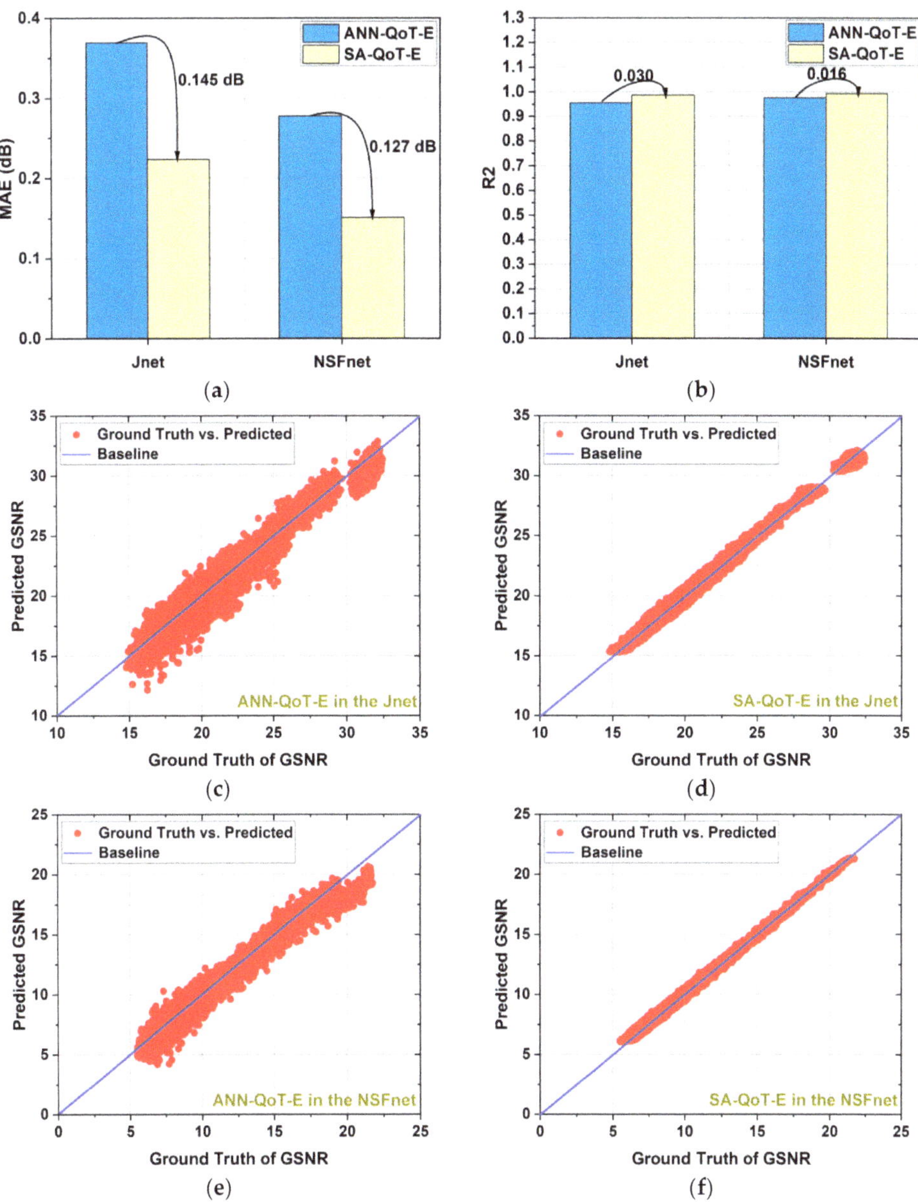

Figure 5. Estimation accuracy of the ANN-QoT-E and the SA-QoT-E: (**a**) testing MAE of two models in the Jnet and NSFnet; (**b**) testing R2 of two models in the Jnet and NSFnet; (**c**) ground truth of GSNR vs. predicted GSNR based on the ANN-QoT-E in the Jnet; (**d**) ground truth of GSNR vs. predicted GSNR based on the SA-QoT-E in the Jnet; (**e**) ground truth of GSNR vs. predicted GSNR based on the ANN-QoT-E in the NSFnet; (**f**) ground truth of GSNR vs. predicted GSNR based on the SA-QoT-E in the NSFnet.

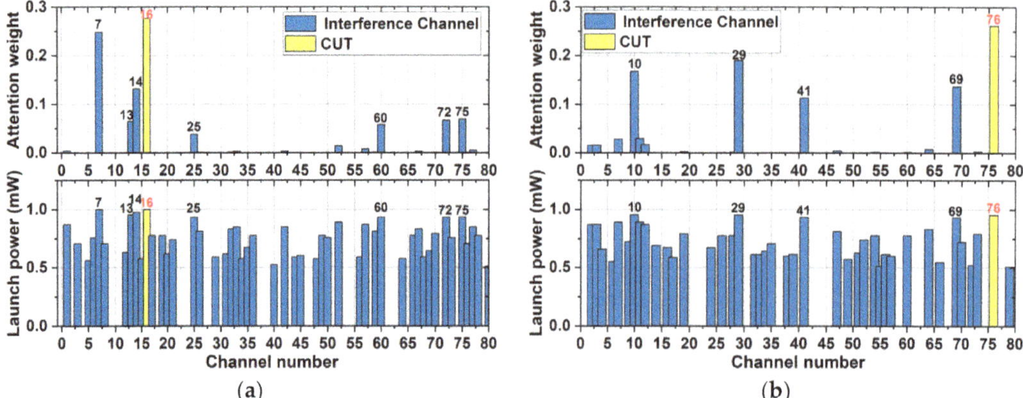

Figure 6. (**a**) Attention weight and launch power of each channel for Ch_16 in the Jnet; (**b**) attention weight and launch power of each channel for Ch_76 in the NSFnet.

4.3. Scalability

The input dimension and output dimension of ANN models are fixed. When the number of network wavelengths is expanded (such as a C band network expanding to a C+L band network), the original ANN-QoT-E cannot be applied to the network. The SA-QoT-E has the advantage of variable length of the input feature vector sequence due to the introduction of the self-attention mechanism. We generate new testing datasets in the Jnet with 120 wavelengths (the full channels of the C++ band), in the Jnet with 216 wavelengths (the full channels of the C+L band), in the NSFnet with 120 wavelengths, and in the NSFnet with 216 wavelengths, which are defined as T_{Jnet}^{120}, T_{Jnet}^{216}, T_{NSFnet}^{120}, and T_{NSFnet}^{216}, respectively. Figure 7 shows the estimation accuracy of the SA-QoT-E in the Jnet and NSFnet obtained by testing on $T_{Jnet}^{80}/T_{Jnet}^{120}/T_{Jnet}^{216}$ and $T_{NSFnet}^{80}/T_{NSFnet}^{120}/T_{NSFnet}^{216}$, where the SA-QoT-E is trained on $D_{Jnet}^{80}/D_{NSFnet}^{80}$. As shown in Figure 7a, in the Jnet, the testing MAEs of the SA-QoT-E tested on $T_{Jnet}^{80}/T_{Jnet}^{120}/T_{Jnet}^{216}$ are low; even the highest MAE obtained on T_{Jnet}^{216} is lower than the MAE obtained by the ANN-QoT-E in Figure 5a; in the NSFnet, the testing MAEs of the SA-QoT-E tested on $T_{NSFnet}^{80}/T_{NSFnet}^{120}/T_{NSFnet}^{216}$ are close and low. Figure 7b shows the R2 score of the SA-QoT-E tested on $T_{Jnet}^{80}/T_{Jnet}^{120}/T_{Jnet}^{216}$ and $T_{NSFnet}^{80}/T_{NSFnet}^{120}/T_{NSFnet}^{216}$, in the Jnet and NSFnet, and the SA-QoT-E achieves a high R2 score.

Figure 7c,d show the predicted value of the SA-QoT-E against their actual GSNR tested on T_{Jnet}^{120} and T_{Jnet}^{216}, respectively. The results show that the SA-QoT-E trained on D_{Jnet}^{80} has a relatively high accuracy on T_{Jnet}^{120} and T_{Jnet}^{216}; specifically, the maximum absolute errors of the SA-QoT-E tested on T_{Jnet}^{120} and T_{Jnet}^{216} are 1.79 dB and 2.58 dB, respectively.

Figure 7e,f show that the SA-QoT-E trained on D_{NSFnet}^{80} has a relatively high accuracy on T_{NSFnet}^{120} and T_{NSFnet}^{216}; specifically, the maximum absolute errors of the SA-QoT-E tested on T_{NSFnet}^{120} and T_{NSFnet}^{216} are 0.84 dB and 0.73 dB, respectively.

The self-attention mechanism learns the interaction between channels, and aggregates multiple channel feature vectors into a new channel feature vector to estimate the QoT of the corresponding channel, Thus, the SA-QoT-E is not limited by the number of network channels. In conclusion, the SA-QoT-E can be directly applied to network wavelength expansion scenarios without retraining.

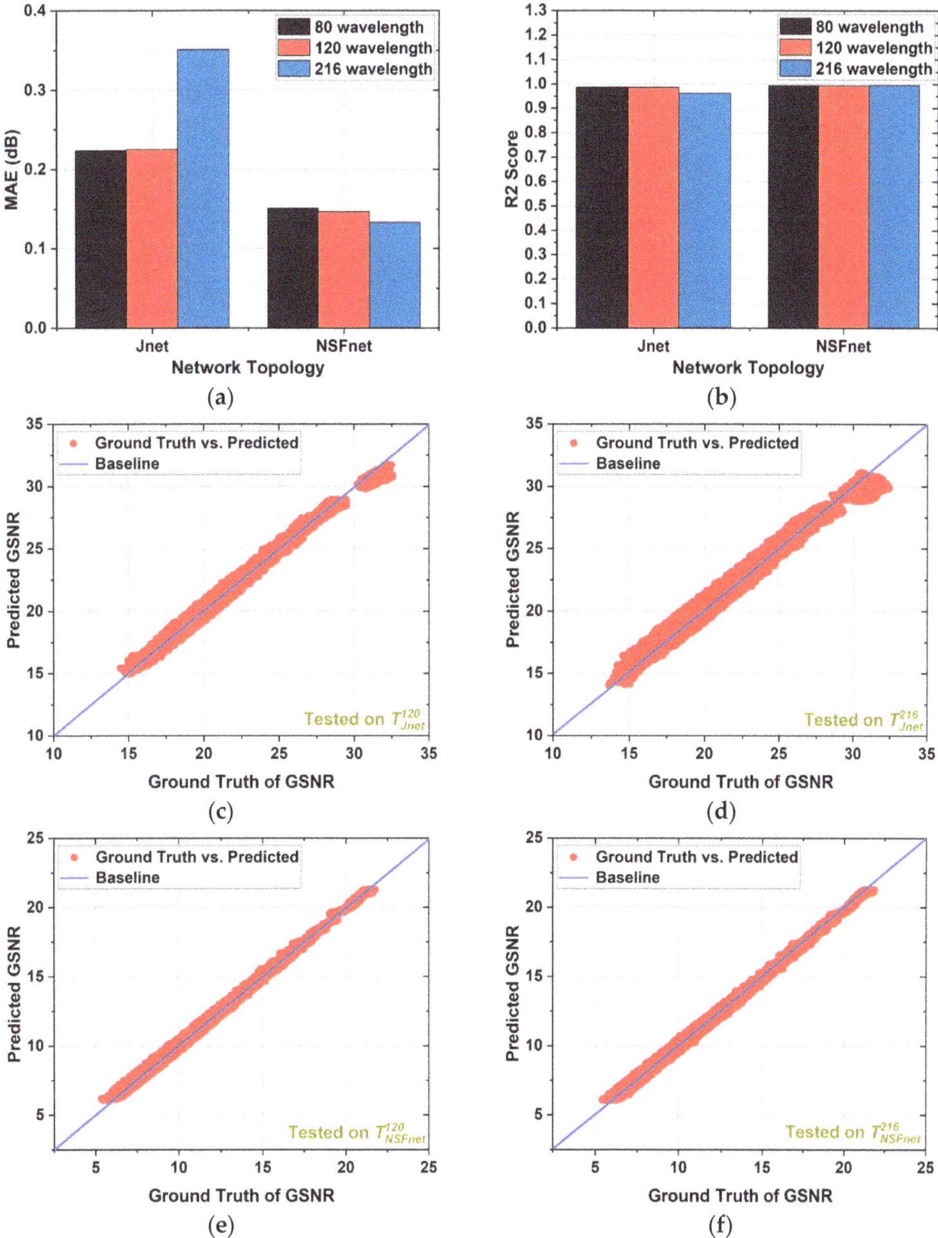

Figure 7. Estimation accuracy of the SA-QoT-E in the Jnet and NSFnet with different numbers of wavelengths: (**a**) testing MAE of the SA-QoT-E in the Jnet and NSFnet; (**b**) testing R2 of the SA-QoT-E in the Jnet and NSFnet; (**c**) ground truth of GSNR vs. predicted GSNR based on the SA-QoT-E in the Jnet with 120 wavelengths; (**d**) ground truth of GSNR vs. predicted GSNR based on the SA-QoT-E in the Jnet with 216 wavelengths; (**e**) ground truth of GSNR vs. predicted GSNR based on the SA-QoT-E in the NSFnet with 120 wavelengths; (**f**) ground truth of GSNR vs. predicted GSNR based on the SA-QoT-E in the NSFnet with 216 wavelengths.

4.4. Computational Complexity

Compared with the ANN-QoT-E, the SA-QoT-E not only improves the estimation accuracy but also has scalability. However, the SA-QoT-E has higher computational complexity. The computational complexity is the total number of addition operations and multiplication operations of the QoT estimators. The computational complexity of each layer of the ANN models is its input dimension multiplied by its output dimension; thus, the computational complexity of the ANN-QoT-E in our simulations can be calculated as follows:

$$C_{ANN} = O\Big((2N_{ch}+1)N_h + N_h^2 + N_h N_{ch}\Big) = O\Big(N_{ch}N_h + N_h^2\Big), \quad (5)$$

where C_{ANN} is the computational complexity of the ANN-QoT-E, N_{ch} is the number of wavelengths in the network, and N_h is the number of neurons in the ANN model's hidden layer.

The self-attention mechanism mainly contains three steps: (1) similarity calculation; (2) soft-max operation; (3) weighted summation. First, the computational complexity of the similarity calculation is $O(N_{ch}^2 n_d)$, due to the fact that it is a $N_{ch} \times n_d$ matrix multiplied by a $n_d \times N_{ch}$ matrix. In addition, the computational complexity of soft-max operation is $O(N_{ch}^2)$. Finally, the weighted summation is a $N_{ch} \times N_{ch}$ matrix multiplied by a $N_{ch} \times n_d$ matrix; thus, its computational complexity is $O(N_{ch}^2 n_d)$. The computational complexity of the ANN model in the SA-QoT-E is $O(N_{ch}(n_d N_h + N_h^2 + N_h))$. Therefore, the computational complexity of the SA-QoT-E is shown as follows:

$$C_{SA} = O\Big(2N_{ch}^2 n_d + N_{ch}^2 + N_{ch}\big(n_d N_h + N_h^2 + N_h\big)\Big) = O\Big(N_{ch}^2 n_d + N_{ch}N_h n_d + N_{ch}N_h^2\Big), \quad (6)$$

where C_{SA} is the computational complexity of the SA-QoT-E. The computational complexity of the ANN-QoT-E and the SA-QoT-E is shown in Table 2. Obviously, the computational complexity of the SA-QoT-E is higher than that of the ANN-QoT-E. However, the training of the estimation models is offline and C_{SA} is acceptable, and the SA-QoT-E can be applied to realistic optical network optimization.

Table 2. Computational complexity of the ANN-QoT-E and the SA-QoT-E.

QoT Estimation Model	Computational Complexity
ANN-QoT-E	$O(N_{ch}N_h + N_h^2)$
SA-QoT-E	$O(N_{ch}^2 n_d + N_{ch}N_h n_d + N_{ch}N_h^2)$

4.5. Discussion

The estimation accuracy of QoT estimators is important to ensure the reliable transmission of the lightpaths in optical networks, and the computational complexity of QoT estimators decides their availability in practical scenarios. The SA-QoT-E has the advantages compared with the ANN-QoT-E: (1) stronger data-fitting ability; (2) higher estimation accuracy; (3) stronger scalability. However, these advantages of SA-QoT-E come at the cost of high computational complexity. The computational complexity of SA-QoT-E is higher than that of the ANN-QoT-E, which is mainly affected by the number of channels N_{ch}. Thus, the ANN-QoT-E is more suitable for optical networks with a large number of network channels and a shortage of computing resources. In most scenarios, the SA-QoT-E has more advantages compared with the ANN-QoT-E.

5. Conclusions

In multi-channel optical networks, due to the nonlinear effect of fibers, the different interfering channel has a different effect on the CUT. The existing ANN-QoT-E ignores the different effects of the interfering channels, which affects its estimation accuracy. To improve the accuracy of the ANN-QoT-E, we propose a novel SA-QoT-E, where the self-attention mechanism assigns attention weights to the interfering channels according to their effects on the CUT. Moreover, we use a hyperparameter search method to optimize the hyperparameters of the SA-QoT-E. The simulation results show that the proposed SA-QoT-E improves the estimation accuracy compared with the ANN-QoT-E. Specifically, compared with the ANN-QoT-E, the testing MAE achieved by the SA-QoT-E is decreased by 0.147 dB and 0.127 dB in the Jnet and NSFnet, respectively; the R2 score achieved by the SA-QoT-E is improved by 0.03 and 0.016 in the Jnet and NSFnet, respectively; the maximal absolute error achieved by the SA-QoT-E is reduced from 4.21 dB to 1.13 dB in the Jnet and from 3.46 dB to 0.84 dB in the NSFnet. Moreover, the SA-QoT-E has scalability, which can be directly applied to network wavelength expansion scenarios without retraining. However, compared with the ANN-QoT-E, the SA-QoT-E has higher computational complexity. Fortunately, the training of the SA-QoT-E is offline and the computational complexity of the SA-QoT-E is acceptable; thus, the proposed SA-QoT-E can be applied to the realistic optical network and provide more accurate lightpath QoT information for optical network optimization.

Author Contributions: Conceptualization, Y.Z. and Z.G.; methodology, Y.Z. and Z.G.; software, Y.Z. and Z.G.; validation, Y.Z., X.H. and Z.G.; formal analysis, J.Z. and Y.D.; investigation, J.Z and R.G.; resources, J.Z. and Y.J.; data curation, J.Z. and Z.G.; writing—original draft preparation, Y.Z. and Z.G.; writing—review and editing, Z.G., J.Z., R.G. and Y.J.; visualization, Y.Z.; supervision, X.H. and Y.D.; project administration, X.H. and Y.D.; funding acquisition, Z.G. and R.G. All authors have read and agreed to the published version of the manuscript.

Funding: This work was supported in part by the National Natural Science Foundation of China (No. 62101058), and in part by the Fund of State Key Laboratory of IPOC (BUPT) (No. IPOC2022ZT11), P.R. China.

Institutional Review Board Statement: Not applicable.

Informed Consent Statement: Not applicable.

Data Availability Statement: The data used to support the findings of this study are available from the corresponding author upon request.

Conflicts of Interest: The authors declare no conflict of interest.

References

1. Ayassi, R.; Triki, A.; Crespi, N.; Minerva, R.; Laye, M. Survey on the use of machine learning for quality of transmission estimation in optical transport networks. *J. Light. Technol.* **2022**, *40*, 5803–5815. [CrossRef]
2. Shao, J.; Liang, X.; Kumar, S. Comparison of Split-Step Fourier Schemes for Simulating Fiber Optic Communication Systems. *IEEE Photonics J.* **2014**, *6*, 7200515.
3. Poggiolini, P. The GN Model of Non-Linear Propagation in Uncompensated Coherent Optical Systems. *J. Light. Technol.* **2012**, *30*, 3857–3879. [CrossRef]
4. Aladin, S.; Tremblay, C. Cognitive Tool for Estimating the QoT of New Lightpaths. In Proceedings of the 2018 Optical Fiber Communications Conference and Exposition (OFC), San Diego, CA, USA, 11–15 March 2018.
5. Rottondi, C.; Barletta, L.; Giusti, A.; Tornatore, M. Machine-learning method for quality of transmission prediction of unestablished lightpaths. *IEEE/OSA J. Opt. Commun. Netw.* **2018**, *10*, A286–A297. [CrossRef]
6. Azzimonti, D.; Rottondi, C.; Tornatore, M. Reducing probes for quality of transmission estimation in optical networks with active learning. *IEEE/OSA J. Opt. Commun. Netw.* **2020**, *12*, A38–A48. [CrossRef]
7. Rottondi, C.; Riccardo, D.M.; Mirko, N.; Alessandro, G.; Andrea, B. On the benefits of domain adaptation techniques for quality of transmission estimation in optical networks. *IEEE/OSA J. Opt. Commun. Netw.* **2021**, *13*, A34–A43. [CrossRef]
8. Liu, C.-Y.; Chen, X.; Proietti, R.; Yoo, S.J.B. Performance studies of evolutionary transfer learning for end-to-end QoT estimation in multi-domain optical networks. *IEEE/OSA J. Opt. Commun. Netw.* **2021**, *13*, B1–B11. [CrossRef]
9. Samadi, P.; Amar, D.; Lepers, C.; Lourdiane, M.; Bergman, K. Quality of transmission prediction with machine learning for dynamic operation of optical WDM networks. In Proceedings of the 2017 European Conference on Optical Communication (ECOC), Gothenburg, Sweden, 17–21 September 2017.
10. Seve, E.; Pesic, J.; Delezoide, C.; Bigo, S.; Pointurier, Y. Learning Process for Reducing Uncertainties on Network Parameters and Design Margins. *IEEE/OSA J. Opt. Commun. Netw.* **2018**, *10*, A298–A306. [CrossRef]
11. Yu, J.; Mo, W.; Huang, Y.K.; Lp, E.; Kilper, D.C. Model transfer of QoT prediction in optical networks based on artificial neural networks. *IEEE/OSA J. Opt. Commun. Netw.* **2019**, *11*, C48–C57. [CrossRef]
12. Mahajan, A.; Christodoulopoulos, K.; Martinez, R.; Spadaro, S.; Muñoz, R. Modeling EDFA gain ripple and filter penalties with machine learning for accurate QoT estimation. *J. Light. Technol.* **2020**, *38*, 2616–2629. [CrossRef]

13. Cho, H.J.; Varughese, S.; Lippiatt, D.; Ralph, S.E. Convolutional recurrent machine learning for OSNR and launch power estimation: A critical assessment. In Proceedings of the 2020 Optical Fiber Communications Conference and Exhibition (OFC), San Diego, CA, USA, 28 March–1 April 2020.
14. Azzimonti, D.; Rottondi, C.; Giusti, A.; Tornatore, M.; Bianco, A. Comparison of domain adaptation and active learning techniques for quality of transmission estimation with small-sized training datasets. *IEEE/OSA J. Opt. Commun. Netw.* **2021**, *13*, A56–A66. [CrossRef]
15. Liu, X.; Lun, H.; Liu, L.; Zhang, Y.; Liu, Y.; Yi, L.; Hu, W.; Zhuge, Q. A Meta-Learning-Assisted Training Framework for Physical Layer Modeling in Optical Networks. *J. Light. Technol.* **2022**, *40*, 2684–2695. [CrossRef]
16. Kruse, L.E.; Kühl, S.; Pachnicke, S. Exact component parameter agnostic QoT estimation using spectral data-driven LSTM in optical networks. In Proceedings of the 2022 Optical Fiber Communications Conference and Exhibition (OFC), San Diego, CA, USA, 7–11 March 2022.
17. Bergk, G.; Shariati, B.; Safari, P.; Fischer, J.K. ML-assisted QoT estimation: A dataset collection and data visualization for dataset quality evaluation. *IEEE/OSA J. Opt. Commun. Netw.* **2022**, *14*, 43–55. [CrossRef]
18. Ayoub, O.; Bianco, A.; Andreoletti, D.; Troia, S.; Giordano, S.; Rottondi, C. On the Application of Explainable Artificial Intelligence to Lightpath QoT Estimation. In Proceedings of the 2022 Optical Fiber Communications Conference and Exhibition (OFC), San Diego, CA, USA, 5–9 March 2022.
19. Ayoub, O.; Andreoletti, D.; Troia, S.; Giordano, S.; Bianco, A.; Rottondi, C. Quantifying Features' Contribution for ML-based Quality-of-Transmission Estimation using Explainable AI. In Proceedings of the 2022 European Conference on Optical Communication (ECOC), Basel, Switzerland, 18–22 September 2022.
20. Allogba, S.; Aladin, S.; Tremblay, C. Machine-learning-based lightpath QoT estimation and forecasting. *J. Light. Technol.* **2022**, *40*, 3115–3127. [CrossRef]
21. Panayiotou, T.; Savva, G.; Shariati, B.; Tomkos, I.; Ellinas, G. Machine Learning for QoT Estimation of Unseen Optical Network States. In Proceedings of the 2019 Optical Fiber Communications Conference and Exhibition (OFC), San Diego, CA, USA, 3–7 March 2019.
22. Safari, P.; Shariati, B.; Bergk, G.; Fischer, J.K. Deep Convolutional Neural Network for Network-wide QoT Estimation. In Proceedings of the 2021 Optical Fiber Communications Conference and Exhibition (OFC), San Francisco, CA, USA, 6–10 June 2021.
23. Gao, Z.; Yan, S.; Zhang, J.; Mascarenhas, M.; Nejabati, R.; Ji, Y.; Simeonidou, D. ANN-based multi-channel QoT-prediction over a 563.4-km field-trial testbed. *J. Light. Technol.* **2020**, *38*, 2646–2655. [CrossRef]
24. Zhou, Y.; Gu, Z.; Zhang, J.; Ji, Y. Attention Mechanism Based Multi-Channel QoT Estimation in Optical Networks. In Proceedings of the 2021 Asia Communications and Photonics Conference (ACP), Shanghai, China, 24–27 October 2021.
25. Vaswani, A.; Shazeer, N.; Parmar, N.; Uszkoreit, J.; Jones, L.; Gomez, A.N.; Kaiser, L.; Polosukhin, I. Attention is all you need. In Proceedings of the 31st Conference on Neural Information Processing Systems (NIPS 2017), Long Beach, CA, USA, 4–9 December 2017.
26. Jain, A.K.; Mao, J.; Mohiuddin, K.M. Artificial neural networks: A tutorial. *Computer* **1996**, *29*, 31–44. [CrossRef]
27. Ferrari, A.; Filer, M.; Balasubramanian, K.; Yin, Y.; Le Rouzic, E.; Kundrat, J.; Grammel, G.; Galimberti, G.; Curri, V. GNPy: An open source application for physical layer aware open optical networks. *IEEE/OSA J. Opt. Commun. Netw.* **2020**, *12*, C31–C40. [CrossRef]

Article

Probabilistically-Shaped DMT for IM-DD Systems with Low-Complexity Fast WHT-Based PDSP

Yi Liu [1], Haimiao Long [2], Ming Chen [2,*], Yun Cheng [1] and Taoyun Zhou [1]

[1] Department of Information Science and Engineering, Hunan University of Humanities, Science and Technology, Loudi 417000, China
[2] School of Physics and Electronics, Hunan Normal University, Changsha 410081, China
* Correspondence: ming.chen@hunnu.edu.cn

Abstract: Transmission capacity and receiver sensitivity of an intensity-modulation direct detection (IM-DD) optical discrete multi-tone (DMT) system can be improved by using the probabilistically shaping (PS) technique. However, different probabilistic distributions will be required owing to the unbalanced signal-to-noise ratio (SNR) among data-carrying subcarriers (SCs) induced by the imperfect frequency response of optical/electrical devices, which can increase the implementation complexity of the PS-DMT transceiver. In this work, different signal pre-processing schemes including pre-equalization, Walsh–Hadamard transform (WHT)-based full data-carrying SCs precoding (FDSP) and fast WHT-based partial data-carrying SCs precoding (PDSP) are investigated for SNR equalization in a short-reach PS-DMT transmission system. After transmission over 50 km single-mode fiber, the experimental results indicated that three pre-processed signals have almost the same generalized mutual information (GMI) performance and receiver sensitivity improvements. The proposed fast WHT-based PDSP scheme may be a good option for the implementation of the PS-DMT transmission systems with a large SC SNR fluctuation regarding computational complexity.

Keywords: intensity-modulation and direct-detection (IM-DD); discrete multi-tone (DMT); probabilistically shaping (PS); full data-carrying SC precoding (FDSP); partial data-carrying SC precoding (PDSP); Walsh–Hadamard transform (WHT)

Citation: Liu, Y.; Long, H.; Chen, M.; Cheng, Y.; Zhou, T. Probabilistically-Shaped DMT for IM-DD Systems with Low-Complexity Fast WHT-Based PDSP. *Photonics* **2022**, *9*, 655. https://doi.org/10.3390/photonics9090655

Received: 28 August 2022
Accepted: 12 September 2022
Published: 15 September 2022

1. Introduction

Intensity-modulation and direct detection (IM-DD) optical discrete multi-tone (DMT) has been widely considered a promising candidate for optical fiber access networks, owing to its high spectral efficiency (SE), robustness against fiber dispersions, and low cost [1–3]. However, current optical access networks cannot support further mobile traffic and keep up with the continuously increased bandwidth demand [4,5]. Thus, one effective scheme named probabilistically shaping (PS) technique has been investigated widely in optical transmission systems due to the improved transmission capacity and receiver sensitivity [6,7]. In [8], the authors proposed a fixed-length matcher named constant composition distribution matching (CCDM), and its output data sequence can obey the same empirical distribution (ED). Moreover, the shaped symbols can be represented by using binary tags and encoded by using forward error correction (FEC) code with preserving the distribution of the shaped symbols. Unfortunately, the practical IM-DD DMT transmission systems suffer the large data-carrying subcarriers (SCs) signal-to-noise ratio (SNR) fluctuation caused by various interferences, such as the data converters-induced clock tone leakage (CTL), imperfect optoelectronic devices-induced nonlinear effect and serious low-pass-like attenuation, etc. As a result, different probabilistic distributions may be required for the implementation of the PS technique, which increases the system's complexity.

In the literature, one classic and effective scheme named the adaptive modulation technique is widely used, which can boost the system's capacity [9]. Besides, a simplified scheme named the pre-emphasis (or pre-equalization) technique was also widely used to effectively compensate for power fading in transmission systems [10]. However, the traditional adaptive loaded-DMT or pre-emphasis technique is channel-dependent and requires the channel state information (CSI) with the reverse link, which is complex and time consuming. Nowadays, a channel-independent precoding scheme is seen as another effective way to compensate for unbalanced impairments. Thus, by employing this technique, only one symbol modulation scheme and one kind of CCMD are required in PS-enabled DMT transmission systems. Some precoding matrices such as the Walsh–Hadamard transform (WHT) matrix [11], constant amplitude zero autocorrelation sequence (CAZAC) matrix [12], and discrete Hartley transform (DHT) matrix [13] have been employed to realize precoding for DMT transmission systems. The precoding technique can also be applied to reduce the peak-to-average power ratio (PAPR) owing to the improvement of the autocorrelation performance of signal symbols [14], and mitigate nonlinear distortions induced by electrical/optical devices. In [15], J. Ma et al. proposed and experimentally demonstrated an orthogonal circular matrix transform (OCT)-based precoding scheme in a short-reach IM-DD transmission system, which can effectively reduce the PS-enabled transceiver implementation complexity. Moreover, the precoding schemes based on seven commonly precoding matrices were comparatively investigated [16], and their computational complexities were theoretically analyzed and compared as well. It also indicated that, compared with other precoding matrices (e.g., OCT, DHT, CAZAC, etc.), there is no need for multiplication operations in WHT precoding, which can be considered as a suitable scheme to compensate unbalanced impairments in an optical IM-DD system regarding the computational complexity.

In the above-mentioned precoding works, all of the data-carrying SCs in DMT or OFDM symbols are used for precoding, and we call this method full data-carrying SC precoding (FDSP). Since the Hermitian symmetry (HS) constraint is required for the inverse fast Fourier transform (IFFT) to obtain the real-valued DMT signal, the number of data-carrying SCs is generally not an integer power of two. In this case, the precoding techniques cannot be directly implemented with the corresponding FFT-based fast algorithms. Moreover, when a large number of SCs are employed for data delivery, the FDSP exhibits high complexity from a hardware implementation point of view. To deal with this issue, the multi-band OCT-based precoding technique was proposed [17]. In addition, a block precoding (BL) scheme, in which the data-carrying SCs are divided into several groups and the number of data-carrying SCs in each group is an integer power of 2, was proposed in [18,19]. However, the difference in equalized SNRs is relatively large among these groups. Compared to the uniform modulation scheme, these methods need to adopt different modulation formats and probabilistic distributions according to the SNRs of groups.

In this work, to further reduce the implementation complexity of the precoding-enabled PS-DMT transceiver, a fast WHT-based partial data-carrying SC precoding (PDSP) technique is proposed and experimentally verified in the optical IM-DD transmission system. The proposed PDSP technique can realize SNR equalization and provide a significant reduction in implementation complexity. We also compare the proposed technique with the FDSP and digital pre-equalization techniques. The rest of this paper is structured as follows. The operation principle of the probabilistically shaping technique, fast WHT-based partial data-carrying SCs pre-coding (PDSP) scheme, and its computational complexity is described in Section 2. The experimental setup and verification are presented in Sections 3 and 4, respectively. The corresponding conclusion is finally summarized in Section 5.

2. Operation Principle

2.1. The Principle of PS Technique

The diagram for the PS technique is schematically plotted in Figure 1. At the transmitter, the upper (or lower) data sequence is separated into two bit streams, which contain V_1

and V_2 bits, respectively. The upper (or lower) data sequence is shaped to non-uniform distributed symbols (U_1) through CCDM, and its corresponding rate can be expressed as $R_{DM} = V_1/U_1$. In this scheme, the non-uniform distribution P_X is applied and can be expressed as [20]

$$P_X(x) = e^{-\lambda|x^2|} \Big/ \sum_{x\in\chi} e^{-\lambda|x^2|} \tag{1}$$

where λ is a rate parameter and these shaped symbols (M-PAM) can be expressed as $\chi = \{\pm 1, \pm 3, \ldots \pm (M-1)\}$. In this work, M is chosen to 8. Subsequently, these shaped symbols are mapped into binary bit sequences with labeling by gray mapping and encoded by low-density parity-check (LDPC) code with the DVB-S.2 standard. Note that a bit-level interleaver is applied in our work to further improve the performance of FEC coding [15]. In the following, 2 one-dimensional M-PAM symbols are mapped in I/Q parts, respectively, to generate a two-dimension M^2-QAM symbol. Thus the information rate (IR) of PS-enabled 64-QAM (bits /QAM symbols), R_{PS}, can be expressed as

$$R_{PS} = 2(1 + R_{DM} - m(1 - R_C)) \tag{2}$$

where R_C is the rate of LDPC code and $m = \log_2 M$.

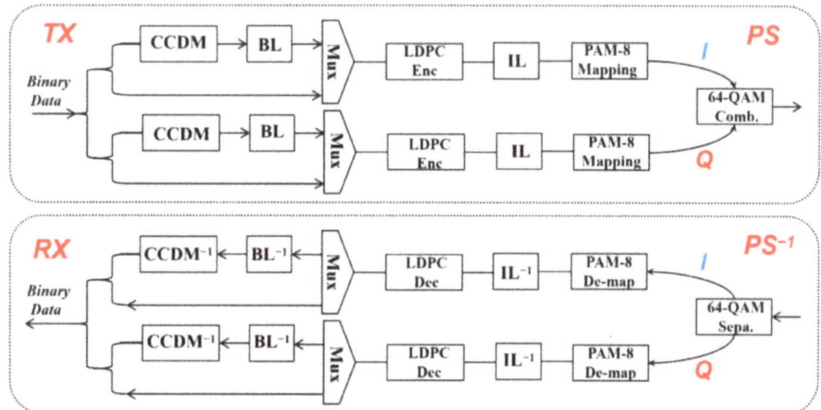

Figure 1. Schematic diagram of the PS-enabled DMT system.

At the receiver, the corresponding inverse operations for the PS technique are also shown in Figure 1. It should be mentioned that the generalized mutual information (GMI) under bit-metric decoding (BMD) is estimated by using log-likelihood ratios (LLRs) [7]. Assuming that N-points samples are obtained with the Monte Carlo simulation method, the GMI for I/Q parts can be written as [21]

$$GMI_{I(Q)} \approx \frac{1}{N}\sum_{k=1}^{N}(-\log_2 P_X(x_k)) - \frac{1}{N}\sum_{k=1}^{N}\sum_{i=1}^{m}\left(\log_2\left(1 + e^{(-1)^{b_{k,i}}L_{k,i}}\right)\right) \tag{3}$$

where $b_{k,i}$ and $L_{k,i}$ represent the input bit and output bit, respectively. According to Equation (3), GMI_I and GMI_Q have the same calculation method, and the total GMI for the M^2-QAM symbol equals to GMI_I + GMI_Q.

2.2. Fast WHT-Based PDSP Scheme

Ideally, only one probability distribution is required for the PS-DMT transmission system with a small range of SC SNR variations. However, the SC SNR may be largely fluctuated due to the imperfect frequency response of the optical/electrical devices, as illustrated in Figure 2a. In the conventional FDSP case, all of the data-carrying SCs are regarded as one group and used for performing precoding (see Figure 2b,c). As mentioned above, the number of data-carrying SCs (F) is usually not an integer power of two due to the HS constraint. Thus, the hardware implementation of the FDSP is a big challenge when a large FFT size is applied for DMT modulation/demodulation. For the proposed PDSP scheme, we divide F data-carrying SCs into two groups: P data-carrying SCs in group 1, where P is a number of an integer power of two, on both sides are selected for precoding, while no precoding is performed for the left F-P data-carrying SCs in group 2, as shown in Figure 2d,e. Note that the P data-carrying SCs selected from the successive data-carrying SCs in high/low-frequency whose SNRs are lower than average SNR and successive low-frequency data-carrying SCs with highest SNRs. In this case, the implementation complexity can be further reduced with the PDSP scheme in the PS-DMT transmission system.

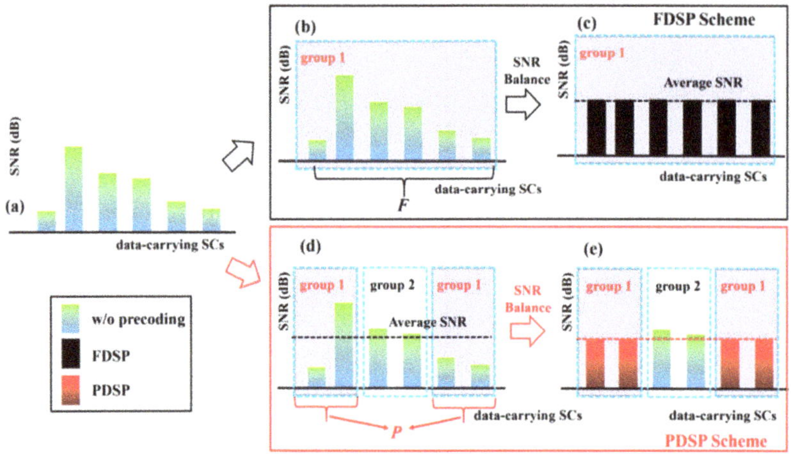

Figure 2. Schematic diagram of the PDSP scheme in the PS-enabled DMT system. (**a**) original SCs SNR, (**b**) SCs selected for FDSP, (**c**) SNR balance for FDSP, (**d**) SCs selected for PDSP and (**e**) SNR balance for PDSP.

Unlike other PDSP schemes with a complex-valued precoded matrix (e.g., DFT and OCT), the real-valued WHT-based PDSP technique has lower computational complexity. According to [16], the normalized Hadamard matrix of order P can be given by

$$D_P = \frac{1}{\sqrt{P}}\begin{bmatrix} D_{P/2} & D_{P/2} \\ D_{P/2} & -D_{P/2} \end{bmatrix}, D_2 = \frac{1}{\sqrt{2}}\begin{bmatrix} 1 & 1 \\ 1 & -1 \end{bmatrix} \quad (4)$$

where D_2 is the normalized Hadamard matrix of order 2, and P must be either 1, 2, or an integer multiple of 4. D_P can be constructed from D_2. In the PDSP scheme, P is an integer number of two, the inverse matrix $D_P{}^{-1}$ equals to D_P.

Once the PS is done, the shaped QAM symbols can be expressed as $\boldsymbol{X}_F = [X_1, X_2, \ldots, X_F]^T$ and $\boldsymbol{X}_P = [X_1, X_2, \ldots, X_P]^T$, where $F > P$ and T denotes the transpose operation. Thus, by multiplying the fast WHT-based precoding matrix D_P, the PDSP precoded symbols $\boldsymbol{Y}_P = [Y_1, Y_2, \ldots, Y_P]^T$ can be given by

$$\boldsymbol{Y}_P = D_P \boldsymbol{X}_P \quad (5)$$

After fiber channel transmission, without considering nonlinear distortions, the recovered PDSP symbols $\boldsymbol{R}_P = [R_1, R_2, \ldots, R_P]$ after FFT operation can be given by

$$\boldsymbol{R}_P = H\boldsymbol{Y}_P + W = HD_P\boldsymbol{X}_P + W \quad (6)$$

The channel transfer matrix, H, is a diagonal matrix that can be expressed as diag $(h_1, h_2, \ldots, h_P)$, and the frequency response of the k-th SC is h_k. Similarly, the noise on the k-th SC is denoted by W_k, which obeys Gaussian distribution with variance and zero mean. The corresponding noise vector in the frequency domain is expressed as $\boldsymbol{W} = [W_1, W_2, \ldots, W_P]$. Assuming that the accurate channel equalization is performed, the PDSP-decoded QAM symbols $\boldsymbol{Z}_P = [Z_1, Z_2, \ldots, Z_P]$ can be written as

$$\boldsymbol{Z}_P = D_P^{-1}H^{-1}\boldsymbol{R}_P = \boldsymbol{X}_P + D_P^{-1}H^{-1}W = \boldsymbol{X}_P + D_P H^{-1}W \tag{7}$$

After performing PDSP decoding, these data-carrying SCs can be treated equally and only one kind of CCDM and one symbol modulation scheme are required for the implementation of the PS-DMT transmission system.

2.3. Complexity Comparison

The computational complexities of FDSP/PSDP schemes are analyzed and listed in Table 1. The number of data-carrying SCs F is not an integer power of two in the WHT-based FDSP scheme, and the required real-valued addition operations may be resource-intensive. However, the fast WHT can be implemented when the number of precoded data-carrying SCs P is an integer power of two. Therefore, the computational complexity of the fast WHT-based PDSP scheme can be significantly reduced.

Table 1. WHT-based FDSP/PDSP computational complexity comparison.

Precoding Scheme	Refer to [16]		Fast Algorithm [18]	
	Real Mult.	Real Add.	Real Mult.	Real Add.
FDSP	0	$2F^2 - 2F$	-	-
PDSP	0	$2P^2 - 2P$	0	$2P\log_2(P)$

3. Experimental Setup

The experimental setup of the PS-DMT systems enabled by a fast WHT-based PDSP scheme is illustrated in Figure 3. At the transmitter (Tx), the digital PS-DMT signal is generated offline with digital signal processing (DSP) approaches in Matlab. Firstly, the pseudo-random binary sequence (PRBS) is generated and then sent to the PS module, which is clearly mentioned in Section 2.1. After that, the shaped 64-QAM symbols are precoded with the proposed fast WHT scheme. Note that only 96 and 64 positive-frequency SCs are, respectively, used for FDSP and PDSP schemes, and other SCs are filled with zeros. After the HS operation, the cyclic prefix (CP) with a length of eight points is added for each 256-point inverse fast Fourier transform (IFFT) output to resist ISI. Additionally, one training symbol (TS) is inserted in the front of each PS-DMT frame to realize both timing synchronization and channel estimation [22]. Finally, the precoded PS-DMT signal is digitally clipped to combat the digital-to-analog converter (DAC)-induced quantization noise [23] and reduce PAPR. The analog electrical DMT signals are converted by using a Tektronix arbitrary waveform generation (AWG, AWG7122C). The corresponding sampling rate and the resolution for D/A conversion are 12-GSa/s and 10-bits, respectively. Note that the average signal power for traditional/FDSP/PDSP PS-enabled DMT signals are set to equal in this experiment. The converted electrical DMT signals are suppressed by a low-pass filter (LPF) and then amplified by a Mini-Circuits electrical amplifier (EA, ZX60-14012L-S+) with a bandwidth of ~6 GHz. The Mach–Zehnder modulator (MZM) works at the quadrature point to minimize the non-linear distortion. The wavelength and output power of the laser diode (LD) are 1550 nm and 11 dBm, respectively. Then, the 2.5 dBm optical PS-DMT signals are coupled into a 50 km single-mode fiber (SMF), and its dispersion and loss are ~17 ps/nm/km and ~0.2 dB/km, respectively. At the receiver (Rx), the optical PS-DMT signal is attenuated by a variable optical attenuator (VOA). An optical coupler (OC) is used to change the received optical power (ROP). Then, the output PS-DMT signal with 90% power is detected by a photodiode (PD). The recovered signal is amplified by an EA and captured by a digital storage oscilloscope (DSO, Lecroy Wavemaster 820-Zi-A, Teledyne LeCroy, Chestnut Ridge, New York, USA). The sampling rate of DSO and resolution for A/D conversion are 40 GSa/s and 8 bits, respectively. The sampled signals are post-processed with the Rx DSP flow, which consists of symbol timing synchronization, remove CP, FFT, ISFA-enhanced channel estimation [23], frequency domain equalization, fast WHT decoding, and PS demodulation. Finally, the different BER or GMI performances are calculated.

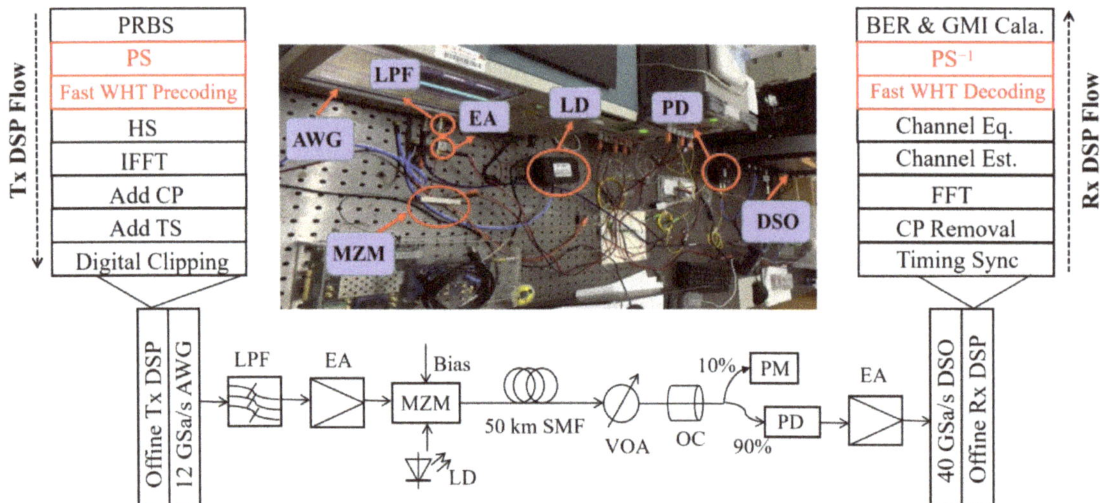

Figure 3. Experimental setup of the PS-DMT system enabled by fast WHT-based PDSP scheme.

When the rate of LDPC is set to 9/10, the different rates of CCDM, i.e., 1.75, 1.55 and 1.25 bits per one-dimensional symbol are discussed in this work. Therefore, according to Equation (2), the corresponding information rates (IRs) are 4.9, 4.5, and 3.9 per two-dimensional symbol (bits/QAM symbol). Meanwhile, the constellations of the three probability distributions for probabilistically shaped 64-QAM are also given in Figure 4. It shows clearly that, as the R_{DM} decreases, the probabilistically shaping for the DMT symbol can become more obvious. The bandwidth of the 64-QAM PS-DMT signal is constantly equal to 4.5 ((96 × 12)/256) GHz. When the IR equal to 4.9 bits/QAM symbol, the corresponding data rate and net data rate are 22.05 (12 × 4.9 × (96/256)) Gb/s and 21.36 (12 × 4.9 × (96/256) (900/901)) Gb/s, respectively. Thus, the achievable spectral efficiency of the PS-DMT signal is 4.74 (21.36/4.5) bit/s/Hz. The key parameters of the precoded PS-DMT frame are indicated in Table 2.

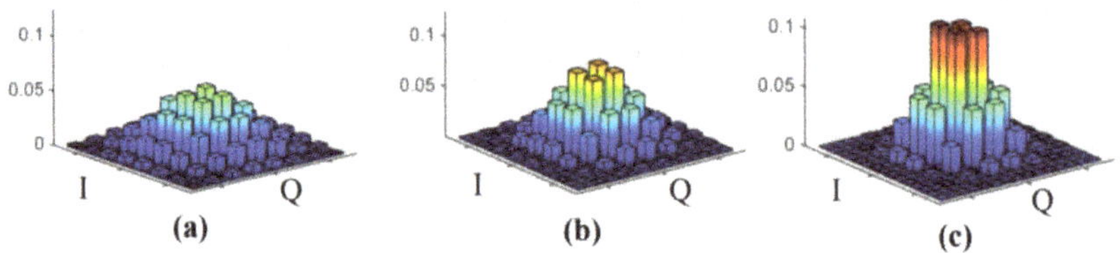

Figure 4. The constellations of the three probability distributions for probabilistically shaped 64-QAM. (**a**) R_{DM} = 1.75, (**b**) R_{DM} = 1.55 and (**c**) R_{DM} = 1.25.

Table 2. Some key parameters of the precoded PS-DMT frame.

Parameter	Value		Unit
	FDSP	PDSP	
Modulation format	64 QAM		-
IFFT/FFT size	256		points
Data SCs	96		-
Data SCs for Precoding	96	64	-
CP length	8		-
TS per frame	1		-
DMT symbols per frame	900		-
Clipping ratio (CR)	13		dB
Bandwidth	4.5		GHz

4. Results and Discussion

The three pre-processing techniques including digital pre-equalization (Preq), WHT-based FDSP and fast WHT-based PDSP are compared for PS-DMT at three IRs regarding electrical spectra, SNR equalization, BER, and GMI performance.

4.1. Spectrum Analysis

The electrical spectra of the transmitted signals for original, digital Preq, WHT-based FDSP and fast WHT-based PDSP schemes are shown in Figure 5a–d, respectively. For the digital Preq scheme, to resist high-frequency attenuation, the mapped QAM symbols are multiplied by a pre-equalization factor via round-trip feedback. Therefore, as shown in Figure 5b, its high-frequency power is relatively high.

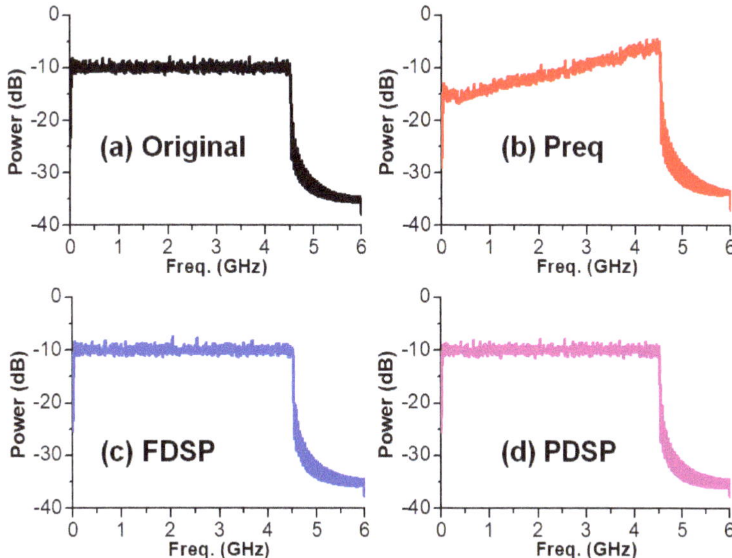

Figure 5. Spectral of the transmitted signals with (**a**) original scheme, (**b**) Preq scheme, (**c**) WHT-based FDSP scheme and (**d**) Fast WHT-based PDSP scheme.

After 50 km SMF transmission, the spectra of the received original, Preq, WHT-based FDSP and fast WHT-based PDSP precoded signals are shown in Figure 6a–d, respectively. Except for the Preq scheme, the high-frequency components of the spectra show obvious attenuation (see Figure 6a,c,d) owing to the low-pass filtering induced by the optical/electrical devices and fiber dispersions. The flat signal spectrum (see Figure 6b) also indicates that the Preq scheme can compensate for the high-frequency attenuation of the PS-DMT signal in the IM-DD transmission system.

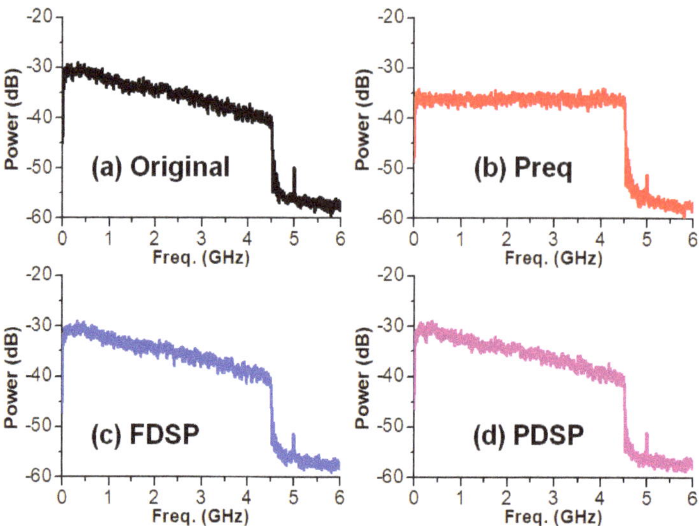

Figure 6. Spectra of the received signals with (**a**) original scheme, (**b**) Preq scheme, (**c**) WHT-based FDSP scheme and (**d**) Fast WHT-based PDSP scheme.

4.2. SNR Equalization

The estimated SC SNRs for four kinds of 64-QAM PS-DMT signals are shown in Figure 7, where the ROP and IR for the DMT symbols are −9 dBm and 4.9 bits/QAM, respectively. The SC SNR fluctuation is up to 13 dB for the original signal. The imperfect frequency response of the MZM and EA are the main reasons for low SNRs on the low-frequency SCs. Additionally, the degraded SNR on the high-frequency SCs is mainly caused by the bandwidth limitations of the optical/electrical devices and fiber dispersion. By using WHT-based FDSP or PDSP schemes, the SNR values on the precoded SCs are well equalized. Unlike the FDSP scheme, only 64 data-carrying SCs, with indices of 1st–32nd and 65th–96th, are used for the proposed PDSP scheme. Therefore, we can observe that the SNR value of the precoded SCs with the PDSP scheme is slightly lower than that of the FDSP scheme, but higher SNR performance on the middle data-carrying SCs with indices from 33rd to 64th is achieved with the PDSP scheme. Compared with the FDSP scheme, the Preq scheme can also play the role of equalizing the SC SNR.

When the ROP is set to −9 dBm, the recovered 64-QAM constellations for four kinds of PS-DMT signals are shown in Figure 8a–d. Compared to the conventional PS-DMT technique, the Preq, WHT-based FDSP and WHT-based PDSP schemes make the constellation points more distinct and convergent.

4.3. GMI and BER Performance

The GMI versus ROP for four different kinds of PS-DMT signal schemes at IRs of 4.9/4.5/3.9 bits/QAM symbol are investigated and shown in Figure 9a–c, respectively. They indicate that Preq, WHT-based FDSP and Fast WHT-based PDSP schemes have almost the same GMI performance and outperform the original one, due to the relatively flat SNR distribution. As the theoretical maximum information rate can be given by GMI with an ideal FEC code, there is about 0.3 dB improvement for the receiver sensitivity in PS-DMT transmission system with the LDPC rate of 9/10.

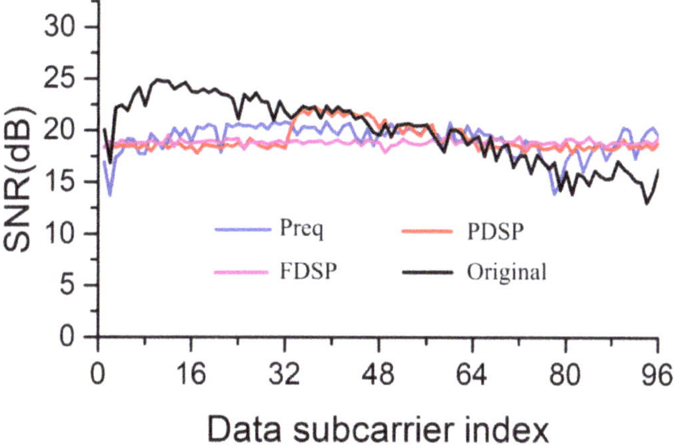

Figure 7. The estimated SC SNR for four kinds of received PS-DMT signals.

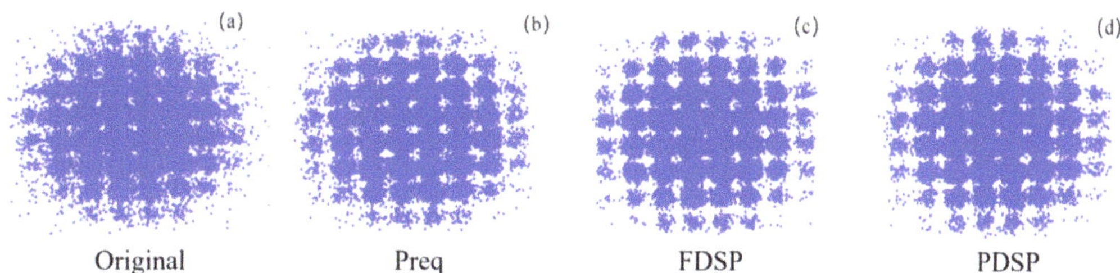

Figure 8. The recovered 64 QAM constellations for (**a**) original scheme, (**b**) pre-equalization, (**c**) WHT-based FDSP and (**d**) fast WHT-based PDSP.

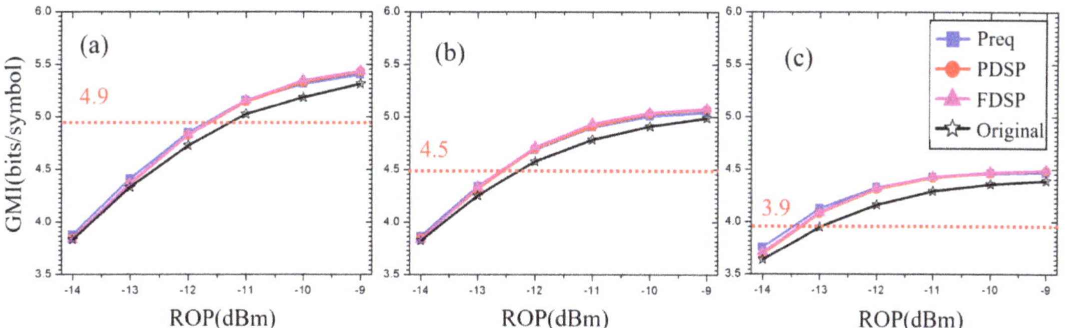

Figure 9. Measured GMI versus ROP for IRs of (**a**) 4.9, (**b**) 4.5 and (**c**) 3.9 bits/QAM symbol.

The measured BER performance for four kinds of PS-DMT signal scheme versus ROP is investigated and shown in Figure 10. When the LDPC rate is set to 9/10 and IRs equal to 4.9/4.5/3.9 bits/QAM symbol, compared with the original scheme, the receiver sensitivity improvements of about 1 dB can be achieved at BER performance below 1e-3 for Preq/WHT-based FDSP/fast WHT-based PDSP schemes. Thus, the fast WHT-based PDSP scheme may be a good option for the implementation of the PS-DMT transmission systems with a large SC SNR fluctuation regarding computational complexity.

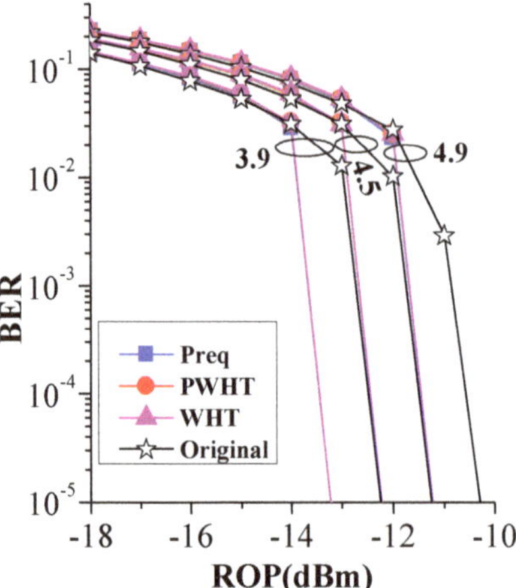

Figure 10. Measured BER performance for four kinds of PS-DMT signals versus ROP.

5. Conclusions

In this paper, we proposed and experimentally investigated a low-complexity fast WHT-based PDSP scheme to combat the unbalanced impairments and reduce the implementation complexity of PS-enabled DMT IMDD transmission systems. After 50 km SMF transmission, the results show that pre-equalization, WHT-based FDSP and fast WHT-based PDSP schemes have almost the same BER and GMI performance. Compared with the traditional PS-enabled scheme, there is about 1 dB gain for receiver sensitivity by employing the three pre-processing schemes. However, the proposed fast WHT-based PDSP scheme is a lower-complexity option for the implementation of the PS-DMT systems with large SC SNR fluctuation.

Author Contributions: Conceptualization, Y.L. and H.L.; methodology, Y.L., H.L. and M.C.; software and validation, Y.L. and H.L.; writing—original draft preparation, Y.L. and H.L.; review and editing, M.C. and T.Z.; supervision, M.C.; project administration, M.C. and Y.C.; funding acquisition, Y.C. All authors have read and agreed to the published version of the manuscript.

Funding: This work is supported by the Scientific Research Fund of Hunan Provincial Education Department of China under grants 20B330 and 21A0562, and the Construct Program of the Key Discipline in Hunan Province, China.

Institutional Review Board Statement: Not applicable.

Informed Consent Statement: Not applicable.

Data Availability Statement: Not applicable.

Conflicts of Interest: The authors declare no conflict of interest. The funders had no role in the design of the study; in the collection, analyses, or interpretation of data; in the writing of the manuscript, or in the decision to publish the results.

References

1. Nguyen, H.; Huang, S.; Wei, C.; Chuang, C.; Chen, J. 55-Gbps and 30-dB Loss Budget LR-OFDM PON Downstream Enabled by ANN-based Predistortion. In Proceedings of the 2021 Optical Fiber Communications Conference and Exhibition (OFC), San Francisco, CA, USA, 6–10 June 2021. paper M3G.4.
2. Li, F.; Li, X.; Chen, L.; Xia, Y.; Ge, C.; Chen, Y. High-level QAM OFDM system using DML for low-cost short reach optical communications. *IEEE Photon. Technol. Lett.* **2014**, *26*, 941–944.
3. Shi, J.; Zhang, J.; Zhou, Y.; Wang, Y.; Chi, N.; Yu, J. Transmission performance comparison for 100Gb/s PAM-4, CAP-16 and DFT-spread OFDM with direct detection. *J. Lightwave Technol.* **2017**, *35*, 5127–5133. [CrossRef]
4. Mei, J.; Li, K.; Ouyang, A.; Li, K. A profifit maximization scheme with guaranteed quality of service in cloud computing. *IEEE Trans. Comput.* **2015**, *64*, 3064–3078. [CrossRef]

5. Fang, Y.; Yu, J.; Zhang, J.; Chi, N.; Xiao, J.; Chang, G.K. Ultrahigh-capacity access network architecture for mobile data backhaul using integrated W-band wireless and free-space optical links with OAM multiplexing. *Opt. Lett.* **2014**, *39*, 4168–4171. [CrossRef] [PubMed]
6. Buchali, F.; Steiner, F.; Böcherer, G.; Schmalen, L.; Schulte, P.; Idler, W. Rate adaptation and reach increase by probabilistically shaped 64-QAM: An experimental demonstration. *J. Lightwave Technol.* **2016**, *34*, 1599–1609. [CrossRef]
7. Böcherer, G.; Steiner, F.; Schulte, P. Bandwidth efficient and ratematched low-density parity-check coded modulation. *IEEE Trans. Commun.* **2015**, *63*, 4651–4665. [CrossRef]
8. Schulte, P.; Böcherer, G. Constant composition distribution matching. *IEEE Trans. Inf. Theory* **2016**, *62*, 430–434. [CrossRef]
9. Chen, X.; Feng, Z.; Tang, M.; Fu, S.; Liu, D. Performance enhanced DDO-OFDM system with adaptive partitioned precoding and single sideband modulation. *Opt. Express* **2017**, *25*, 23093–23108. [CrossRef] [PubMed]
10. Gao, Y.; Yu, J.; Xiao, J.; Cao, Z.; Li, F.; Chen, L. Direct-Detection Optical OFDM Transmission System With Pre-Emphasis Technique. *J. Lightwave Technol.* **2011**, *29*, 2138–2145. [CrossRef]
11. Deng, R.; He, J.; Chen, M.; Zhou, Y. Experimental demonstration of a real-time gigabit OFDM-VLC system with a cost-efficient precoding scheme. *Opt. Commun.* **2018**, *423*, 69–73. [CrossRef]
12. Feng, Z.; Wu, Q.; Tang, M.; Lin, R.; Wang, R.; Deng, L.; Fu, S.; Shum, P.P.; Liu, D. Dispersion-tolerant DDO-OFDM system and simplified adaptive modulation scheme using CAZAC precoding. *J. Lightwave Technol.* **2016**, *34*, 2743–2751. [CrossRef]
13. Ouyang, X.; Jin, J.; Jin, G.; Zhang, W. Low complexity discrete Hartley transform precoded OFDM for peak power reduction. *Electron. Lett.* **2012**, *48*, 90–91. [CrossRef]
14. Jiang, T.; Tang, M.; Lin, R.; Feng, Z.; Chen, X.; Deng, L.; Fu, S.; Li, X.; Liu, W.; Liu, D. Investigation of DC-biased optical OFDM with precoding matrix for visible light communications: Theory, simulations, and experiments. *IEEE Photonics J.* **2018**, *10*, 7906916. [CrossRef]
15. Ma, J.; He, J.; Wu, K.; Chen, M. Performance enhancement of probabilistically shaped OFDM enabled by precoding technique in an IM-DD system. *J. Lightwave Technol.* **2019**, *37*, 6063–6071. [CrossRef]
16. Chen, M.; Wang, L.; Xi, D.S.; Zhang, L.; Zhou, H.; Chen, Q.H. Comparison of different precoding techniques for unbalanced impairments compensation in short-reach DMT transmission system. *J. Lightwave Technol.* **2020**, *38*, 6202–6213. [CrossRef]
17. Hong, Y.; Xu, J.; Yeh, L.K.C.C.H.; Liu, L.Y.; Chow, C.W. Experimental investigation of multi-band OCT precoding for OFDM-based visible light communications. *Opt. Express* **2017**, *25*, 12908–12914. [CrossRef] [PubMed]
18. Wang, L.; Chen, M.; Chen, G.; Deng, A.; Zhou, H.; Liu, Y.; Cheng, Y. Fast WHT-based block precoding for DMT transmission. In Proceedings of the Asia Communications and Photonics Conference, Shanghai, China, 24–27 October 2021. paper W4B.6.
19. Li, F.; Xiao, X.; Li, X.; Dong, Z. Real-Time Demonstration of DMT-based DDO-OFDM Transmission and Reception at 50 Gb/s. In Proceedings of the 39th European Conference and Exhibition on Optical Communication (ECOC 2013), London, UK, 22–26 September 2013. paper 1–3.
20. Cho, J.; Winzer, P.J. Probabilistic constellation shaping for optical fiber communications. *J. Lightwave Technol.* **2019**, *37*, 1590–1607. [CrossRef]
21. Fehenberger, T.; Alvarado, A.; Böcherer, G.; Hanik, N. On probabilistic shaping of quadrature amplitude modulation for the nonlinear fiber channel. *J. Lightwave Technol.* **2016**, *34*, 5063–5073. [CrossRef]
22. Liu, Y.; He, J.; Chen, M.; Xiao, Y.Q.; Cheng, Y. 64APSK Constellation Scheme for Short-Reach DMT with ISDD Enabled SFO Compensation. *Opt. Commun.* **2020**, *467*, 125689. [CrossRef]
23. Chen, M.; He, J.; Fan, Q.; Dong, Z.; Chen, L. Experimental Demonstration of Real-Time High-Level QAM-Encoded Direct-Dection Optical OFDM Systems. *J. Lightwave Technol.* **2015**, *33*, 4632–4639. [CrossRef]

MDPI

Article

Low-Complexity and Highly-Robust Chromatic Dispersion Estimation for Faster-than-Nyquist Coherent Optical Systems

Tao Yang [1], Yu Jiang [1], Yongben Wang [2], Jialin You [1], Liqian Wang [1] and Xue Chen [1,*]

[1] State Key Lab of Information Photonics and Optical Communications, Beijing University of Posts and Telecommunications, Beijing 100876, China
[2] ZTE Corporation, Shenzhen 518055, China
* Correspondence: xuechen@bupt.edu.cn

Abstract: Faster-than-Nyquist (FTN) coherent optical transmission technology is considered to be an outstanding solution to achieve higher spectral efficiency (SE), larger capacity, and greater achievable transmission by using advanced modulation formats in concert with highly efficient digital signal processing (DSP) to estimate and compensate various impairments. However, severe inter-symbol interference (ISI) caused by tight FTN pulse shaping will lead to intractable chromatic dispersion (CD) estimation problems, as existing conventional methods are completely ineffective or exhibit unaffordable computational complexity (CC). In this paper, we propose a low-complexity and highly robust scheme that could realize accurate and reliable CD estimation (CDE) based on a designed training sequence (TS) in the first stage and an optimized fractional Fourier transform (FrFT) in the second stage. The training sequence with the designed structure helps us to estimate CD roughly but reliably, and it further facilitates the FrFT in the second stage to achieve accurate CDE within a narrowed searching range; it thereby results in very low CC. Comprehensive simulation results of triple-carrier 64-GBaud FTN dual-polarization 16-ary quadrature amplitude modulation (DP-16QAM) systems demonstrate that, with only overall 3% computational complexity compared with conventional blind CDE methods, the proposed scheme exhibits a CDE accuracy better than 65 ps/nm even under an acceleration factor as low as 0.85. In addition, 60-GBaud FTN DP quadrature phase shift keying (DP-QPSK)/16QAM transmission experiments are carried out, and the results show that the CDE error is less than 70 ps/nm. The advantages of the proposed scheme make it a preferable candidate for CDE in practical FTN coherent optical systems.

Keywords: faster-than-Nyquist; chromatic dispersion estimation; low-complexity

Citation: Yang, T.; Jiang, Y.; Wang, Y.; You, J.; Wang, L.; Chen, X. Low-Complexity and Highly-Robust Chromatic Dispersion Estimation for Faster-than-Nyquist Coherent Optical Systems. *Photonics* **2022**, *9*, 657. https://doi.org/10.3390/photonics9090657

Received: 10 August 2022
Accepted: 12 September 2022
Published: 15 September 2022

1. Introduction

With the exponential growth of data traffic due to bandwidth-intensive applications such as HD video streaming, cloud computing, automatic driving, 5G and other emerging applications, the optical fiber communication capacity is gradually approaching the Shannon limit nowadays. The future optical fiber network is developing towards an ultra-high spectral efficiency, ultra-large capacity and ultra-long transmission distance [1,2]. Optical transport adopting higher bit rates and better spectrum efficiency (SE) is an inevitable step for next-generation wavelength division multiplex (WDM) systems. Although large-capacity WDM transmission systems have been commercially deployed due to the advances of digital signal processing (DSP), further increasing the spectral efficiency is a key challenge in order to meet the explosive demand for higher capacity in communication channels. The mainstream Nyquist WDM system is a feasible scheme to achieve high-spectral-efficiency and large-capacity optical transmission to a certain extent by compressing signal bandwidth and reducing the channel spacing. However, the orthogonal transmission criterion limits the further improvement of the system's spectral efficiency. In recent years, faster-than-Nyquist (FTN) optical transmission technology has drawn great attention and been widely investigated [3–5]. By compressing the symbol interval in the time domain or the channel spacing in the frequency domain, the channel spacing in an FTN system could be less than the symbol rate, and the transmission rate could be higher than the Nyquist rate. Accordingly, FTN optical coherent systems are reasonably considered as a promising solution to achieve a higher SE, larger capacity and greater achievable transmission by using high-order quadrature amplitude modulation (QAM) formats in concert with advanced DSP to compensate the inherent inter-symbol interference (ISI) [6]. Therefore, FTN technology is one of the potential development directions in the field of ultra-high-spectral efficiency and ultra-large-capacity coherent optical transmission.

Chromatic dispersion (CD) compensation is a vital part of digital signal processing (DSP) units for future FTN optical transmissions to exempt from the complicated link dispersion management in the optical domain. CD is also an important optical characteristic of optical fibers, and with the increase of the speed and distance of optical transmission systems, it has become a main obstacle for the development of ultra-high-speed and ultra-long-distance optical transmission systems. In addition, CD, as one of the main channel damages, will cause serious ISI unavoidably in long-haul communication systems. Besides this, due to the real-time requirement of protection switching in the actual system, it is essential for DSP to quickly and accurately monitor the link CD and perform precise compensation. Consequently, a command of accurate CD value in advance is the necessary prerequisite for CD compensation through the DSP algorithms. Moreover, due to the CD estimation (CDE) in the first step of the DSP flow in a digital coherent receiver, at this time, other impairments have not been compensated/equalized, such that the CD estimation algorithm must be tolerant to various types of other damage, thereby putting forward high demands on the robustness of the CD estimation method. Most problematic of all is the FTN strong filtering brought about by introducing serious ISI, which may lead to unaffordable computational complexity (CC) in the estimation of CD, and may further exacerbate the difficulty of CD estimation.

Various CD estimation schemes in a digital coherent receiver have been proposed and demonstrated. One approach relies on extracting parameters from the equalizer taps, but it only applies to the case of a small amount of CD [7]. In order to improve the range of CDE, other dispersion estimation methods are developed. Many CDE schemes based on searching have been reported. One is based on the Godard Clock-Tone (GCT) of the signal spectrum [8,9]. In this method, the power of the clock tone after each signal equalization is calculated by scanning the fixed CD in a pre-defined CD range. The value of the CD can be obtained when the power of the clock tone is the largest. However, it fails when an acceleration factor is small. Another method is based on the peak-to-average-power ratio (PAPR), which also requires a pre-defined CD range and a given CD search step in order to obtain the real CD value. The eigenfunction values need to be calculated after each signal equalization, and the value of CD can be obtained when the power of the eigenfunction is the smallest [10,11]. The scheme based on auto-correlation functions of signal power waveform [12,13] can obtain the CD by finding the time delay when the signal power auto-correlation function is the maximum value, but the scheme completely fails in an FTN-WDM system. Blind CDE based on fractional Fourier transform (FrFT) uses a linear frequency modulation (LFM) signal and has good energy accumulation characteristics with the optimal fraction order [14–17]. The CD can be obtained when the optimal order is found. However, the method is almost the same as the CDE based on PAPR, with both looking up the CD value primarily by little-step searching, and thereby resulting in very high CC. Moreover, the data-aided CDE method was proposed in [18], but the CDE error is too large to be tolerated in FTN-WDM systems. A CDE scheme based on TS was also proposed in [19]. The scheme is used to solve the problem that the blind CDE algorithm based on FrFT fails at ultra-low sampling rates. After equivalent sampling, it still needs the blind CDE based on FrFT to search the optimal fraction order, thereby resulting in high CC. Besides this, the CDE method based on constant amplitude zero autocorrelation (CAZAC) TS in [20] makes use of the similar characteristics of CAZAC TS and a chirp signal. The CDE value can be obtained by using the relationship between the CD and the optimal order. However, it needs to obtain the optimal orders of the TS before and after transmission by scanning. Therefore, the CC of the method is also high. A CDE scheme based on machine learning has high CDE accuracy [21]. However, the CDE range is too small to meet requirement of FTN-WDM system for a long transmission distance. The delay-tap-sampling-based CDE method proposed in [22] needs to establish a CD lookup table, which needs to scan CD with a small step, which also leads to high CC.

In this paper, a low-complexity and highly-robust CDE scheme is proposed and demonstrated. The first stage CDE of the proposed CDE scheme mainly relies on channel estimation using a designed TS, and the fine estimation in the second stage employs blind CDE based on an optimized FrFT. The feasibility of the proposed scheme is verified by the simulation and experiment of 64/60GBaud dual polarization 16-ary QAM (DP-16QAM) and dual polarization quadrature phase shift keying (DP-QPSK) systems under different transmission distances. The results show that the CD estimation performance is accurate and robust. Compared with the conventional blind CDE scheme based on FrFT, the complexity of the proposed scheme is reduced by about 97%, with estimation accuracy comparable to the blind CDE scheme. The remainder of the paper is structured as follows. The principle of the proposed CDE scheme and the complexity comparison are described in Section 2. In Section 3, simulation results are presented to prove the feasibility of the scheme. In Section 4, we confirm the feasibility of the algorithm by performing the experiment. The conclusions are drawn in Section 5.

2. Principle of the CDE Scheme

In conventional scanning/searching-based CDE schemes, we need a lot of samples and a small search step in order to ensure the estimation accuracy; thus, it results in a very high CC that seriously hinders its practical application. Especially in the FTN system, the strong filter damage makes the CD estimation extremely difficult, and it is urgent to study and design a low-complexity and high-robust CDE scheme. Accordingly, here, we propose a novel two-stage CDE scheme. Figure 1 shows the structure of the data frame, which consists of CDE TS used in the first stage and the valid transmission service signal. The notation * means taking the complex conjugate. The TSs are pairs of "a[n]" and "b[n]" sequences which both are CAZAC Chu [19,20] sequences with the same length. Due to the good autocorrelation and randomness of CAZAC Chu sequences, they are chosen as the transport matrix to estimate the channel phase response corresponding to CD. The approximate dispersion value of the system is obtained by channel estimation. Then, according to the range of coarse CDE at the first stage, we further modified and optimized the blind CDE based on FrFT for fine estimation in the second stage. The steps of the CD estimation are shown in Figure 2.

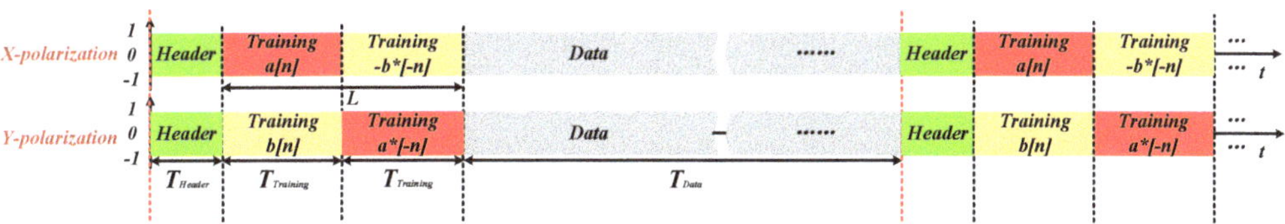

Figure 1. Frame structure diagram.

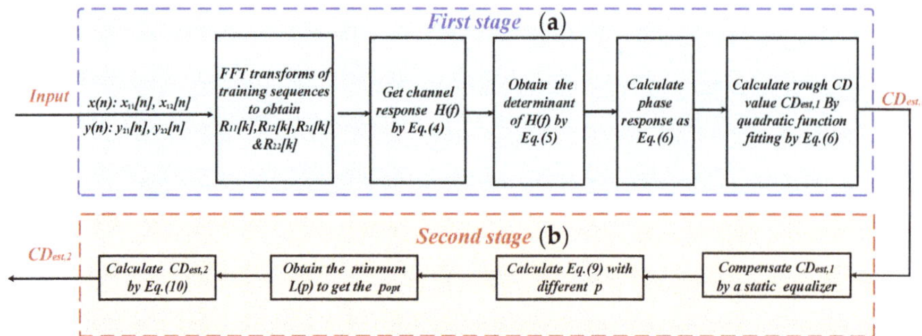

Figure 2. Block diagram of the proposed CDE scheme consisting of two stages: (**a**) the first coarse stage using TSs, and (**b**) the second fine stage using optimized FrFT.

From Figure 2, we can see the principle of the first-stage, where $x_{11}[n]$, $x_{12}[n]$, $y_{21}[n]$ and $y_{22}[n]$ are the input signals that are compensated by the IQ imbalance recovery algorithm corresponding to the TS a[n], −b*[−n], b[n] and a*[−n]. The channel transmission matrix can be expanded as Equation (1), where $H(f)$ represents the channel response of the transmission system, E denotes the polarization-dependent loss (PDL) matrix, $H_{CD}(f)$ is the transfer function of the dispersion information, and $P(f)$ denotes the polarization-mode dispersion (PMD) matrix:

$$H(f) = E \cdot H_{CD}(f) \cdot P(f) \tag{1}$$

where $H_{CD}(f)$ can be described as an all-pass filter in the frequency domain characterized by Equation (2). Here, λ_0 is the central wavelength of the laser source, D is the fiber dispersion parameter, z is the total length of the transmission link, and c is the speed of light:

$$H_{CD}(f) = exp\left(-jf^2 \frac{\pi \lambda_0{}^2 Dz}{c}\right) \tag{2}$$

In transmission systems, because there is mainly the amplitude loss from PDL, the PDL damage represented as E can be ignored reasonably. The PMD transfer matrix $P(f)$ can be written as Equation (3), where $\Delta\tau$ is the time delay caused by PMD, and 2θ and φ are the azimuth and elevation rotation angels, respectively:

$$P(f) = \begin{pmatrix} p(f) & q(f) \\ -q^*(f) & p^*(f) \end{pmatrix} = \begin{pmatrix} e^{j\pi f \Delta\tau} cos\theta & e^{j\varphi} sin\theta \\ -e^{-j\varphi} sin\theta & e^{-j\pi f \Delta\tau} cos\theta \end{pmatrix} \tag{3}$$

where the relationship between $p(f)$ and $q(f)$ can be expressed as $|p(f)|^2 + |q(f)|^2 = 1$. As such, the channel response $H(f)$ of the transmission system can be expressed as Equation (4):

$$H(f) = H_{CD}(f) \begin{pmatrix} p(f) & q(f) \\ -q^*(f) & p^*(f) \end{pmatrix} \tag{4}$$

The determinant of $H(f)$ is as follows:

$$det(H(f)) = exp\left(2jf^2 \frac{\pi \lambda_0{}^2 Dz}{c}\right)\left(|p(f)|^2 + |q(f)|^2\right) = exp\left(2jf^2 \frac{\pi \lambda_0{}^2 Dz}{c}\right) \tag{5}$$

Equation (5) shows that the phase angle depends on CD, such that the coarse CD value Dz can be calculated by the quadratic function simulation graph derived from data fitting, as follows:

$$CD_{est,1} = Dz = \frac{c}{f^2 \pi \lambda_0{}^2} arg\sqrt{det(H(f))} \tag{6}$$

However, the accuracy of the first-stage CDE algorithm cannot meet the requirements of the FTN system; therefore, the second-stage fine CDE algorithm is essential. The principle is as follows: when the linear frequency modulation (LFM) signal has the FrFT transformation of the best order, the maximum energy convergence can be obtained [16]. A similar character can be extended to the signal with CD in the frequency domain by following Equation (7). $F(z, f)$ is the signal with CD in the frequency domain [17].

$$F(z,f) = F(0,f)exp\left(-j\frac{\pi f^2 {\lambda_0}^2 Dz}{c}\right) \tag{7}$$

The optimal order p_{opt} can be inferred as follows:

$$p_{opt} = \frac{2}{\pi} arccot\left(\frac{{\lambda_0}^2 Dz}{c(N-1)\Delta t^2}\right) \tag{8}$$

According to the time–frequency distribution of an optical signal with CD, N represents the number of samples to be calculated and Δt represents the symbol sampling interval. We use p, which ranges from -1 to 1, as the granularity of the search order, and $L(p)$ represents the convergence degree of the signal energy, as shown in Equation (9). Thus, the value of p_{opt} can be obtained when the power of $L(p)$ is at its maximum.

$$L(p) = \int_{-\infty}^{+\infty} \left|X_p(u)\right|^2 du \tag{9}$$

where $Xp(u)$ is the FrFT of the chirped signal in the time domain with the p order. $CD_{est,2}$ is the estimated value we obtain. Finally, the fine CDE expression can be realized as follows:

$$CD_{est,2} = \frac{c(N-1)\Delta t^2}{{\lambda_0}^2} tan\left(p_{opt}\frac{\pi}{2}\right) \tag{10}$$

Complexity Comparison with the Blind CD Estimation Method

We compare the CC in terms of the number of computing operations (mainly the real multiplication and real addition) of the proposed CDE scheme with a conventional blind CDE algorithm based on FrFT [17]. It should be pointed out that all of the complexities are calculated for two polarizations, and the computation is based on optimum implementations. The CC is averaged over a single output sampling symbol. Thus, the CC of the proposed scheme is summarized as follows.

For the first-stage CDE of our proposed scheme, (1) the number of the single-polarization samples is $2L$. Therefore, the FFT operation takes $4\log_2 2L$ real multipliers, and $2\log_2 2L$ real adders. (2) The real multipliers and real adders required by division in the frequency domain for channel estimation, respectively, are 22 and 18. For the second-stage CDE, because the first-stage CDE narrows down the CDE range, the order search range is reduced to $[-k, k]$. (1) The step size of the scanning order is p, and the number of samples is N, which requires FrFT calculation. The CC of FrFT is the same as FFT with the same length. In this stage, $\frac{4k \log_2 N}{p}$ real multipliers and $\frac{2k \log_2 N}{p}$ real adders are needed. Under the same estimation accuracy, the conventional blind CDE algorithm based on FrFT, the order search range is within $(-1, 1)$ and the step of the scanning order is also p, the sample number is N, and the search times are $2/p$. The algorithm takes $\frac{4 \log_2 N}{p}$ real multipliers and $\frac{2 \log_2 N}{p}$ real adders, respectively. In general, the value of L is small, such as 128 in a 10-Gbaud DP-16QAM system [19]. Besides this, because the first-stage CDE narrows down the CDE range, the range of p goes down, and then K is much less than 1. Consequently, the overall CC of the proposed algorithm could be far less than the conventional CDE based on FrFT. The complexity comparison between the two CDE schemes is summarized as shown in Table 1. Because the key parameters of the algorithms will be determined by simulations, the specific complexity comparison will be given in the third section of the simulation verification. For more comprehensive comparison, here we can make a rough estimation of the CC of the proposed scheme by referring to typical parameters in the existing literature. A data-aided CDE scheme is proposed in [19], where the CDE error is about 10% of the real CD. This shows that the first-stage CDE based on TS could narrow down the CDE range markedly, thereby reducing the scanning times by approximately 90%. In the literature [20], 4096 samples and the step size 0.001 of the scanning order are used to search the optimal order of the energy concentration function. It can be inferred from the above typical parameters that the length of the TS used for coarse CDE is usually less than 512. Thus, according to the calculation of the CC in Table 1, we can show that when the CDE accuracy is almost the same, the CC of the proposed CDE scheme is about 3% of the traditional CDE scheme.

Table 1. Complexity comparison between the proposed and conventional CDE scheme.

Algorithm	Real Multiplication	Real Addition
Conventional CDE based on FrFT	$\frac{4\ \log_2 N}{p}$	$\frac{2\ \log_2 N}{p}$
Proposed two-stage CDE	$4\ \log_2 2L + \frac{4k\cdot \log_2 N}{p} + 22$	$2\ \log_2 2L + \frac{2k\cdot \log_2 N}{p} + 18$

3. Simulation and Discussion

3.1. Simulation Environment

In order to verify the feasibility of the proposed CDE algorithm, a three-wavelength 64GBaud DP-16QAM FTN-WDM coherent optical transmission simulation platform was built. In the transmitter-side offline DSP, 2^{19}-1 pseudorandom bit sequences (PRBS) are mapped to 16QAM symbols. The TS is inserted as shown in Figure 1. Then, FTN filtering shaping is completed by Root Raised Cosine (RRC) digital filter with a roll-off factor of 0.01. The transmitter uses three lasers with a channel spacing of 75 GHz as the carrier of the IQ modulator. The central carrier wavelength is 1550 nm. The three modulated optical carriers enter the optical fiber transmission through the wavelength division multiplexing coupler, and the optical fiber link is composed of multiple spans using a fiber loop; each loop includes 80 km optical fiber, a ROADM and an EDFA. At the receiver, the wavelength division multiplexing de-multiplexer is used to filter out the target center wavelength channel. After passing through the optical coherent receiver, the signal is processed by offline DSP to recover the original signals. Before the CD estimation, IQ quadrature imbalance compensation must be performed. In the system, the acceleration factor α is defined as the ratio between the RRC 3 dB bandwidth and the signal baud rate. The fiber CD coefficient and PMD coefficient are 16 ps/nm/km and $0.2\text{ps}/\sqrt{\text{km}}$, respectively. The optical fiber transmission distance is over 160~960 km. The launch power at the transmitter is 0 dBm.

Figure 3 shows a simulation platform for a 64Gbaud DP-16QAM FTN-WDM system and the proposed CD estimation scheme. The parameters of the conventional blind CDE algorithm are the same as the proposed scheme. We use p for the step of the search order. For each CDE, the difference between the real CD value (CD_{real}) and the estimated CD value (CD_{est}) is used to evaluate the performance. M is the total number of computations. The mean error of the absolute CDE (ME_{CD}) is defined as

$$ME_{CD} = \frac{1}{M}\sum_{i=1}^{M}\left|CD_{est,i} - CD_{real,i}\right| \tag{11}$$

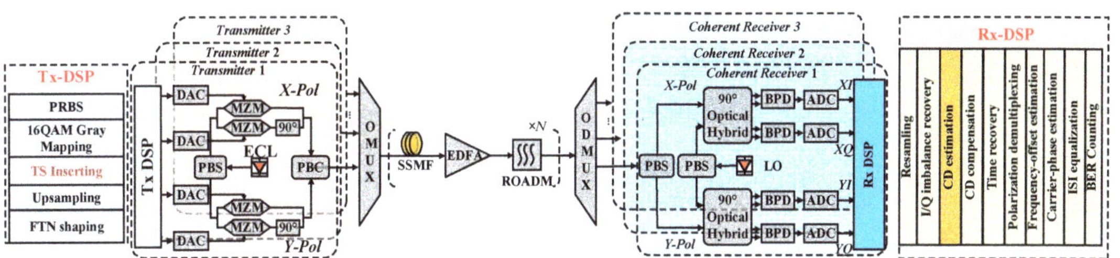

Figure 3. Simulation platform for a 64Gbaud DP-16QAM FTN-WDM system with the proposed CD estimation scheme.

3.2. Analysis of the Impact of the Algorithm Key Parameter

Obviously, the performance of the channel estimation depends on the length of the TS. To some extent, the longer the sequence used to fit the channel phase response, the better the fitting effect will be. Therefore, CD_{est} will be more close to practical CD_{real}. However, excessively long training sequences lead to excessive overhead that will decrease the data transmission efficiency. On the contrary, when the length of TS is short, the estimation will not be accurate enough. It is of great significance to find out the most suitable parameters of the CD estimation algorithm considering the performance and overhead costs.

Theoretically, the length of a single training sequence is required when the training sequence is used to accurately detect CD, as follows [19]:

$$L = Dz{\lambda_0}^2 BN_t/cT \tag{12}$$

where B represents the signal Baud rate and N_t = 2 when the system has two polarizations, and T is the sampling period. In these systems of around 1000-km fiber transmission, the minimum length of TS samples is calculated as 2100. Because the overall CDE accuracy depends on the estimation accuracy of the second-stage, it is enough for L to be a value of 512 in order to obtain a coarse CD in the first stage. Meanwhile, the CDE range of the second-stage can be reduced to about 1/8 of the maximum CD. This indicates that the CDE in the second stage only needs to meet the requirement of a high accuracy within the CD range from −2000 ps/nm to 2000 ps/nm.

Obviously, from the basic principle of the algorithm, we can find that the sample number used in the second fine-stage CDE and the step of the search order p both influence the final CDE accuracy. Due to the first stage of CDE, the search range of the second fine-stage CDE is reduced to [−2000–2000] ps/nm. The range of order p varies with the number of samples. In order to fully estimate the residual CD after the first-stage CDE and compensation, we set the range to be slightly larger than [−2000–2000] ps/nm for the second-stage CDE. Therefore, when the sample numbers are 4096 and 8192, we can obtain from Equation (10) that the search ranges of the order are [−0.048–0.048] and [−0.024–0.024], respectively.

In order to balance the estimation performance and computational complexity of the algorithm, the granularity of the CD scanning, calculated from the sample number and the step of the search order, is set as 100 ps/nm and 200 ps/nm, respectively. Thus, we used 4096 and 8192 samples to evaluate the performance of CDE, and the step size p of the order search is set as 0.002 and 0.004, and 0.001 and 0.002, respectively. The CD_{real} is randomly generated from 160 to 960 km, and each configuration is evaluated 100 times.

Figure 4a,b show the distribution of the mean value of the absolute CDE error using 4096 samples when the step size of the order scanning is 0.002/0.004 and the granularity of the CD scanning is 100/200 ps/nm, respectively. It can be seen that the smaller the CD scanning granularity, the smaller the mean value of the CD estimation error, which is mainly concentrated in the range of less than 65 ps/nm. When the CD scanning granularity is 200 ps/nm, the maximum CD estimation error reaches 115 ps/nm, and the estimation error distribution is relatively uniform. In order to achieve higher CD estimation performance, we increase the number of samples to 8192 for CD estimation. The results in Figure 5 show that the CD estimation performance using 8192 samples is slightly better than that using 4096 samples, and the CD estimation error distributions of the two cases are basically similar.

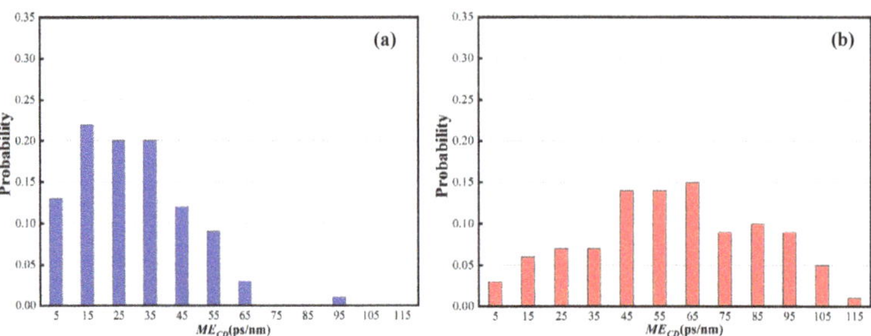

Figure 4. Distribution of the mean value of the absolute CDE error using 4096 samples when the step size of the order scanning and the granularity of the CD scanning are (**a**) 0.002 and 100 ps/nm, and (**b**) 0.004 and 200 ps/nm.

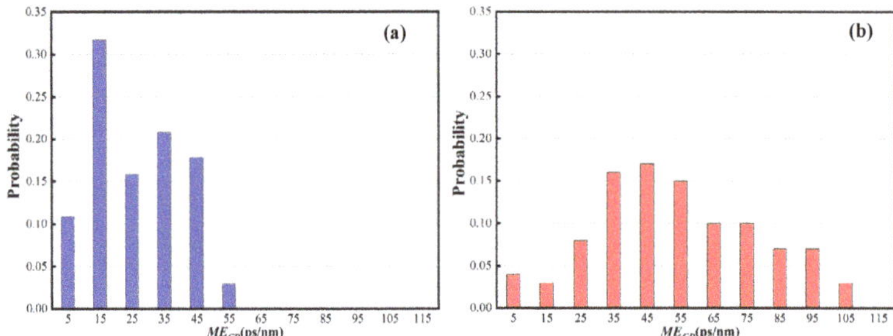

Figure 5. Distribution of the mean value of the absolute CDE error using 8192 samples when the step size of the order scanning and the granularity of the CD scanning are (**a**) 0.001 and 100 ps/nm, and (**b**) 0.002 and 200 ps/nm.

In Table 2, ME_{CD} is the absolute mean error of CDE, and R_{CD} represents the range of the CDE error. We can see that the performance of CDE is similar when scanning the CD with the same granularity. This also indicates that the more sampling points there are, the higher the accuracy of the CDE. With the increase of the step of CD scanning, the effect of CDE becomes worse. When taking the overall complexity of the algorithm proposed into consideration, we select a granularity of CD scanning of 200 ps/nm, and p is 0.004 and 0.002. Besides this, the computational complexity of the two cases with different sample numbers is nearly equivalent. However, in multiple simulations of various link conditions, we found that the CD estimation performance with the sample number of 8192 is more stable; that is, the ability to resist various system impairments (such as PMD, timing error, phase noise, etc.) is stronger. Therefore, considering the ability and stability, the number of sampling symbols is selected as 8192, corresponding to the p of 0.002. In addition, according to the analysis of the algorithm complexity, with the same parameter, i.e., 8192 sampling symbols and a 0.002 step size of order scanning, the overall CC of the proposed CDE algorithm is reduced by about 97% compared with the conventional, completely blind CDE scheme. In addition, as shown in Table 2, we can also compare the CC under different parameters with different CDE performances. When the number of sampling symbols and p are 4096 and 0.002, respectively, the overall CC of the proposed CDE algorithm is reduced by about 95%. If p is 0.004, the overall CC of the proposed CDE algorithm is reduced by about 94.7%. When the number of sampling symbols and p are 8192 and 0.001, respectively, the overall CC of the proposed CDE algorithm is reduced by about 97.5%. Therefore, the proposed CDE algorithm benefits from the two-stage structure, which greatly reduces the CC compared with the traditional algorithm.

Table 2. Performance of the CD estimation.

Sample	p	Granularity 100 ps/nm		Sample	p	Granularity 200 ps/nm	
		ME_{CD}	R_{CD}			ME_{CD}	R_{CD}
4096	0.002	29	2~95	4096	0.004	59	1~114
8192	0.001	26	1~57	8192	0.002	54	0~109

3.3. Evaluation of the CD Estimation Performance

Firstly, in order to verify the tolerance of the ASE noise of the CDE scheme, we set different OSNR values and test the mean error of the CDE for 16QAM, when other system parameters are kept constant. The standard single-mode fiber (SSMF) transmission length is 960 km here. We also set p as 0.002, 0.008, 0.01, and 0.012 in order to fully evaluate the performance of the conventional blind CDE algorithm. The steps of CD scanning are about 200, 800, 1000 and 1200 ps/nm, respectively. Figure 6 shows the simulation results for CD estimation performance against different levels of OSNR for 16-QAM systems using the proposed two-stage scheme. It can be seen that a higher OSNR has a slightly positive impact on the estimation performance.

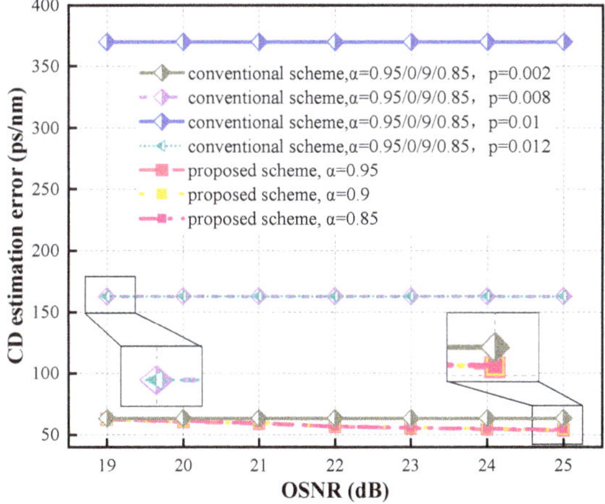

Figure 6. The mean value of the absolute CDE error vs. different OSNRs.

When the FTN acceleration factor α is greater than or equal to 0.85, the CDE performance decreases slightly with the decrease of α, but the difference between the conventional scheme and the proposed one is very small. Overall, the proposed CDE scheme exhibits a CDE accuracy better than 65 ps/nm. Theoretically, for the conventional scheme, the relationship between CD_{est} and p is based on a set trigonometric function instead of a linear function such as formula (10). Besides this, accumulated CD is definite, and the principle of the algorithm makes the searchable CD a discrete value, such that there is a theoretical CDE deviation value. It can be calculated that the CDE errors are about 40, 160, 360 and 160 ps/nm, and we can see that the simulation results are consistent with the theory. Moreover, under the same CDE parameters, i.e., p = 0.002, the accuracy of the conventional blind CDE algorithm based on FrFT is almost equivalent to the CDE we proposed. At this time, the overall CC is reduced by about 97%. With the same fixed number of samples, the computational complexity of the blind CDE algorithm is reduced by increasing p. Even so, the computational complexity of the proposed CDE algorithm is still reduced by at least 85%, while the proposed method has better performance than the conventional one.

In addition, in order to further explore how different transmission distances, i.e., different accumulated CD, affect the performance of the CDE method, the simulation for CD estimation performance against different levels of CD is carried out in a 64Gbaud DP-16QAM FTN-WDM system. Here, accumulated CD is the product of the transmission distance and the CD coefficient of 16 ps/nm/km. Theoretically, the first-stage CDE only determines the residual CD of the second-stage within a certain range that is not fixed. Therefore, as long as the accuracy of the second-stage estimation is high enough, the CDE error of our scheme should be random and relatively small, and the error should basically remain at the same level under different transmission distances. The simulation result proves the above analysis. It can be seen from Figure 7 that the performance of the CDE algorithm is very robust even when the acceleration factor is 0.85. The ME_{CD} is about 55 ps/nm. However, for the conventional blind CDE, the accumulated CD of transmission in one circle is 1280 ps/nm, which is not integer times of the CD search granularity, such that the estimated error varies periodically under such a 200 ps/nm CD search granularity. This is because with the fixed CD searching granularity, the CD estimation performance is very stable even when the acceleration factor is as low as 0.85. If the CD search granularity becomes very small, the differences among different α will become apparent with larger CC. On the contrary, benefitting from the two-stage structure of the proposed algorithm, the proposed algorithm exhibits more stable and accurate CD estimation performance.

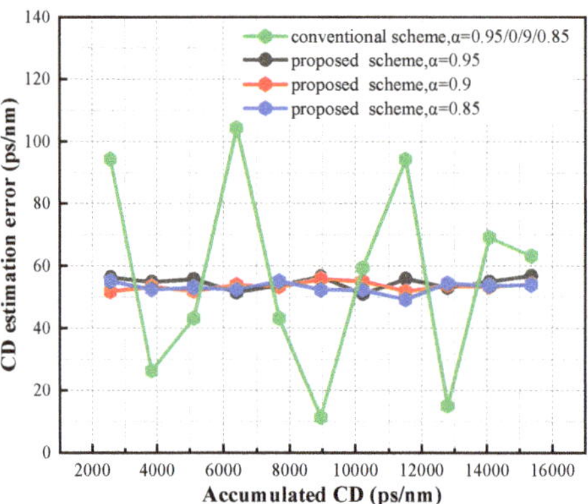

Figure 7. The mean value of the absolute CDE error vs. different CD_{real}.

As we all know, differential group delay (DGD) is one of the important link impairments affecting signal quality, especially in a high-order QAM system. Here, we also simulate the average error of CDE under different DGD in to explore the effect of DGD on the performance of the CDE method. The residual CD and DGD can be compensated using an adaptive equalization algorithm based on a multi-tap finite impulse response (FIR) filter. The results show that the performance of the proposed CDE algorithm decreases with the increase of DGD regardless of the transmission distance in Figure 8. When the DGD is small, the CD estimation performance is mainly determined by the algorithm parameters, and the influence of the acceleration factor α is very small, such that the performance of the CDE is basically the same. However, as the transmission distance increases, the DGD increases, and the ASE noise on the link becomes more and more serious; thereby, the CD estimation is worse for long-distance transmission than for short transmissions. Besides, in the case of large DGD, due to the small acceleration factor, the ISI of the FTN signal is severe, and the CD estimation performance is relatively degraded. Taking the 130 pm/nm residual CD that the FIR equalizer can tolerate as the threshold, the minimum tolerable DGD of the proposed algorithm is about 27 ps.

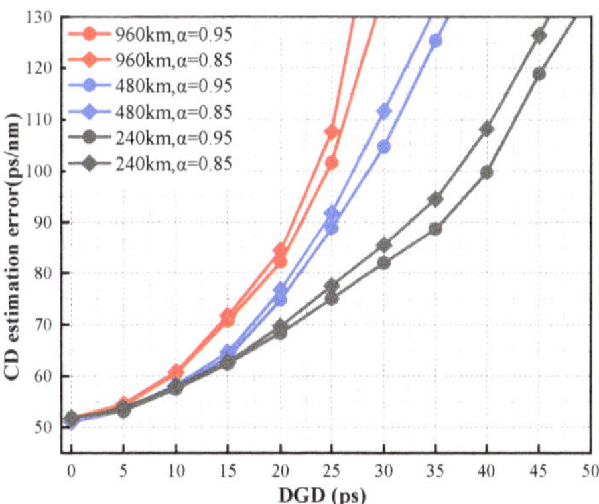

Figure 8. The mean value of the absolute CDE error vs. different DGD.

4. Experimental Setup and Performance Analysis

In order to further verify the practical performance of the proposed CDE algorithm, multi-span fiber transmission experiments were carried out. Figure 9 shows the experimental setup of 60-GBaud DP-QPSK and DP-16QAM coherent optical systems. In the transmitter-side offline DSP, firstly, 2^{19}-1 pseudorandom bit sequences (PRBS) were generated and then mapped into Gray-mapped DP-QPSK/16QAM symbols. The TS was inserted at the transmitter side, as shown in Figure 1. Note that the FEC encoding was not employed in the experiment, and FTN pulse shaping was realized in the digital domain by FIR filters here. Then, the discrete signals after FTN filtering were sent into four synchronized channels of an arbitrary waveform generator (AWG, Agilent 8194A) working at 120 GSa/s with a ~36-GHz 3-dB analog bandwidth. Double oversampling was performed to generate 60Gbaud electrical signals, and tunable external cavity lasers (ECL) with a ~50 kHz measured linewidth were used as the carrier laser. The AWG output differential electrical signals drove a DP I/Q modulator. Then, the modulated carrier was launched into a re-circulating transmission fiber loop, which consisted of a timing controller, an optical coupler, an acousto-optic modulator (AOM)-based optical switch, a fiber span of 80 km SSMF with an attenuation of 0.17 dB/km and CD of 16 ps/nm/km, and an EDFA. Here no inline or pre/post-optical CD compensation were used in the experiment. Note that due to experimental equipment limitation, the shortest distance supported by the fiber loop was 80 × 2 km. After each fiber loop transmission, FTN DP-QPSK/16QAM signals were send to a programmable optical band pass filter (OBPF) with 0.5 nm 3-dB bandwidth and 1 GHz resolution. At the receiver side, another ECL with ~50-kHz linewidth was used as the LO in order to realize coherent detection. The FTN DP-QPSK/16QAM optical signal was sampled using an 80-GSa/s real-time sampling oscilloscope after photoelectric conversion for offline processing. The offline DSP flow is also shown in Figure 9. Firstly, it lowered the sample to two samples per symbol and compensated for IQ imbalance. Then, the proposed and referenced CDE algorithms were implemented before CD compensation. Next, after timing recovery, a radius-directed equalization (RDE) algorithm was employed for the purpose of ISI pre-equalization and polarization de-multiplexing. The frequency offset estimation and carrier phase estimation were subsequently carried out before performing the ISI mitigation based on the post-filter + MLSE scheme. Finally, more than 10^6 symbols were used for BER counting.

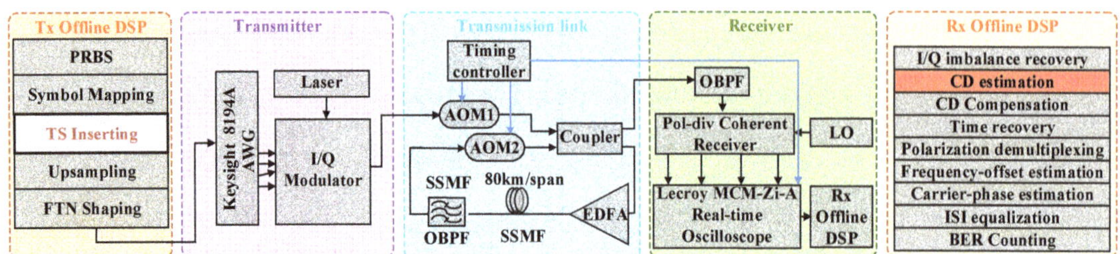

Figure 9. Experimental setup for 60-GBaud FTN DP-QPSK/16QAM coherent transmission systems.

Figure 10 shows the total accumulated CD versus the estimated CD under different FTN acceleration factors α, in the case of an 160–720 km SSMF optical fiber transmission distance. It was observed that both the proposed CDE and conventional blind scheme exhibit substantially equivalent performance. However, because the FTN coherent optical system was extremely sensitive to residual CD, the large CD estimation error will cause the subsequent DSP algorithm to fail to work. Therefore, it was necessary to further quantitatively analyze the error of CD estimation.

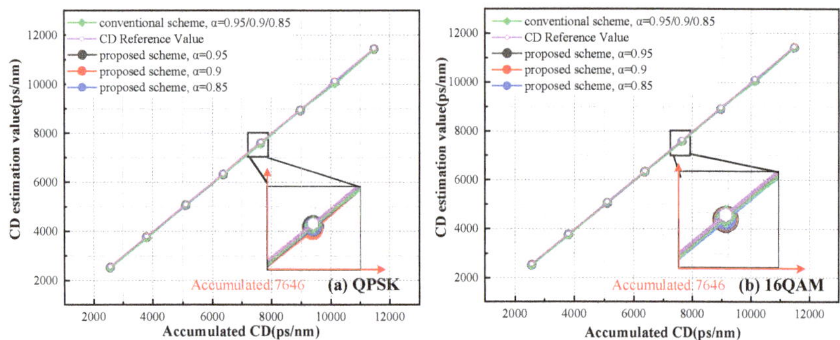

Figure 10. Accumulated CD vs. the CD estimated in (**a**) DP-QPSK and (**b**) DP-16QAM systems.

Figure 11a,b show the mean value of the absolute CDE error in a 60Gbaud DP-QPSK and DP-16QAM experimental system with various link lengths. Although the CDE accuracy fluctuates, it is in the overall range of 40–70 ps/nm. The CDE performance of the 16QAM system was slightly lower than that of the QPSK system, but the difference was small, and the performance has no obvious correlation when α is greater than or equal to 0.85. Compared with the traditional blind CDE algorithm, the CD estimation performance of the proposed algorithm was more stable in both QPSK and 16QAM systems. In addition, as is consistent with the simulation, the CDE error of the blind CDE algorithm varied periodically with the CD_{real} value. Therefore, our proposed CDE algorithm has better estimation accuracy and more robust working stability under different modulation formats with different transmission distances and acceleration factors.

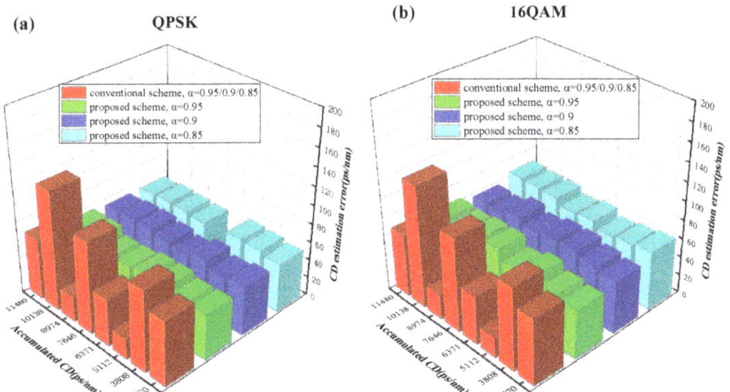

Figure 11. (**a**) The mean value of the absolute CDE error in (**a**) DP-QPSK and (**b**) DP-16QAM systems.

5. Conclusions

In this paper, a low-complexity and highly robust CD estimation scheme for FTN signals was proposed and demonstrated. Taking advantage of a structure combining a designed training sequence with an improved FrFT method, our proposed scheme could estimate CD roughly but reliably in the first stage, and could further facilitate accurate CDE in second stage with very low CC. Complexity analysis showed that, compared with the conventional blind CDE scheme based on FrFT, the overall CC of the proposed scheme is reduced by about 97%, whilst it still maintains better CD estimation performance. Comprehensive simulation results of triple-carrier 64-GBaud FTN-WDM systems show that the CDE accuracy of the proposed scheme is 65 ps/nm over various transmission distances with CD ranges of 2560~15,360 ps/nm for 64-GBaud DP-16QAM signals. Besides this, the long-haul transmission simulation results show that it can tolerate a DGD of at least 27 ps. Furthermore, 60-GBaud DP-QPSK and DP-16QAM signals were generated in experiments for transmission over various distances, and the worst CD estimation errors were 60 ps/nm and 70 ps/nm for QPSK and the 16QAM system, further proving that compared with the conventional CDE methods, only 256 training symbols and overall 3% CC are needed in order to achieve more accurate and more robust CD estimation no matter the different modulation formats, different acceleration factors, and different CD ranges. Therefore, with its low complexity and strong robustness, this CD estimation scheme is a preferable candidate for practical FTN coherent optical systems.

Author Contributions: Conceptualization, T.Y. and X.C.; formal analysis, T.Y. and Y.J.; investigation, T.Y., Y.J. and Y.W.; data curation, T.Y. and X.C.; writing—original draft preparation, T.Y., Y.J. and J.Y.; writing—review and editing, T.Y., L.W. and X.C.; visualization, T.Y. and Y.J.; supervision, T.Y., Y.J., Y.W., J.Y., L.W. and X.C.; project administration, T.Y. and X.C. All authors have read and agreed to the published version of the manuscript.

Funding: This work is partly supported by the National Natural Science Foundation of China (62001045), Beijing Municipal Natural Science Foundation (4214059), the Fund of State Key Laboratory of IPOC (BUPT) (IPOC2021ZT17), and Fundamental Research Funds for the Central Universities (2022RC09).

Institutional Review Board Statement: Not applicable.

Informed Consent Statement: Not applicable.

Data Availability Statement: The data presented in this study are available on request from the corresponding author. The data are not publicly available because the data also form part of an ongoing study.

Acknowledgments: The authors express their appreciation to the reviewers for their valuable suggestions.

Conflicts of Interest: The authors declare no conflict of interest.

References

1. Zhou, G.; Jiang, T.; Xiang, Y.; Tang, M. FrFT based blind chromatic dispersion estimation mitigating large DGD induced uncertainty. In Proceedings of the Asia Communications and Photonics Conference 2020, Beijing, China, 24–27 October 2020; p. M4A.305.
2. Jiang, H. Joint Time/Frequency Synchronization and Chromatic Dispersion Estimation with Low Complexity Based on a Superimposed FrFT Training Sequence. *IEEE Photonics J.* **2018**, *10*, 1–10.
3. Wang, Y.; Liang, J.; Zhou, W.; Wang, W. Accurate chromatic dispersion estimation for Nyquist and Faster Than Nyquist Coherent Optical Systems. In Proceedings of the Asia Communications and Photonics Conference 2021, Shanghai, China, 24–27 October 2021; p. T4A.61.
4. An, S.; Li, J.; Li, X.; Su, Y. FTN SSB 16-QAM Signal Transmission and Direct Detection Based on Tomlinson-Harashima Precoding with Computed Coefficients. *J. Lightwave Technol.* **2021**, *39*, 2059–2066. [CrossRef]
5. Li, S.; Yuan, W.; Yuan, J.; Bai, B.; Wing Kwan Ng, D.; Hanzo, L. Time-Domain, vs. Frequency-Domain Equalization for FTN Signaling. *IEEE Trans. Veh. Technol.* **2020**, *69*, 9174–9179. [CrossRef]
6. Ibrahim, A.; Bedeer, E.; Yanikomeroglu, H. A Novel Low Complexity Faster-than-Nyquist (FTN) Signaling Detector for Ultra High-Order QAM. *IEEE Open J. Commun. Soc.* **2021**, *2*, 2566–2580. [CrossRef]
7. Hauske, F.N.; Geyer, J.; Kuschnerov, M.; Piyawanno, K.; Lankl, B.; Duthel, T.; Fludger, C.R.S.; van den Borne, D.; Schmidt, E.-D.; Spinnler, B. Optical performance monitoring from FIR filter coefficients in coherent receivers. In Proceedings of the Optical Fiber Communication Conference, San Diego, CA, USA, 24–28 February 2008; p. OThW2.
8. Hauske, F.N.; Zhang, Z.; Li, C.; Xie, C.; Xiong, Q. Precise, robust and least complexity CD estimation. In Proceedings of the Optical Fiber Communication Conference, Los Angeles, CA, USA, 6–10 March 2011; p. JWA032.
9. Malouin, C.; Thomas, P.; Zhang, B.; O'Neil, J.; Schmidt, T. Natural Expression of the Best-Match Search Godard Clock-Tone Algorithm for Blind Chromatic Dispersion Estimation in Digital Coherent Receivers. In Proceedings of the Photonics Congress, Colorado Springs, CO, USA, 19–21 June 2012; p. SpTh2B.4.
10. Xie, C. Chromatic Dispersion Estimation for Single-Carrier Coherent Optical Communications. *IEEE Photonics Technol. Lett.* **2013**, *25*, 992–995. [CrossRef]
11. Ospina, R.S.B.; dos Santos, L.F.D.; Mello, A.A.; Ferreira, F.M. Scanning-Based Chromatic Dispersion Estimation in Mode-Multiplexed Optical Systems. In Proceedings of the 21st International Conference on Transparent Optical Networks (ICTON), Angers, France, 9–13 July 2019; pp. 1–4.
12. Pereira, F.C.; Rozental, V.N.; Camera, M. Experimental Analysis of the Power Auto-Correlation-Based Chromatic Dispersion Estimation Method. *IEEE Photonics J.* **2013**, *5*, 7901608. [CrossRef]
13. Sui, Q.; Lau, A.P.T.; Lu, C. Fast and Robust Blind Chromatic Dispersion Estimation Using Auto-Correlation of Signal Power Waveform for Digital Coherent Systems. *J. Lightwave Technol.* **2013**, *31*, 306–312. [CrossRef]
14. Zhou, H. A fast and robust blind chromatic dispersion estimation based on fractional fourier transformation. In Proceedings of the European Conference on Optical Communication (ECOC), Valencia, Spain, 27 September–1 October 2015; pp. 1–3.
15. Sun, X.; Zuo, Z.; Su, S.; Tan, X. Blind Chromatic Dispersion Estimation Using Fractional Fourier Transformation in Coherent Optical Communications. In Proceedings of the IEEE International Conference on Artificial Intelligence and Information Systems (ICAIIS), Dalian, China, 20–22 March 2020; pp. 339–342.
16. Zhou, H. Fractional Fourier Transformation-Based Blind Chromatic Dispersion Estimation for Coherent Optical Communications. *J. Lightwave Technol.* **2016**, *34*, 2371–2380. [CrossRef]
17. Wang, W.; Qiao, Y.; Yang, A.; Guo, P. A Novel Noise-Insensitive Chromatic Dispersion Estimation Method Based on Fractional Fourier Transform of LFM Signals. *IEEE Photonics J.* **2017**, *9*, 1–12. [CrossRef]
18. Do, C.C. Data-Aided Chromatic Dispersion Estimation for Polarization Multiplexed Optical Systems. *IEEE Photonics J.* **2012**, *4*, 2037–2049. [CrossRef]
19. Ma, Y. Training Sequence-Based Chromatic Dispersion Estimation with Ultra-Low Sampling Rate for Optical Fiber Communication Systems. *IEEE Photonics J.* **2018**, *10*, 1–9. [CrossRef]
20. Wu, F.; Guo, P.; Yang, A.; Qiao, Y. Chromatic Dispersion Estimation Based on CAZAC Sequence for Optical Fiber Communication Systems. *IEEE Access* **2019**, *7*, 139388–139393. [CrossRef]
21. Li, J.; Wang, D.; Zhang, M. Low-Complexity Adaptive Chromatic Dispersion Estimation Scheme Using Machine Learning for Coherent Long-Reach Passive Optical Networks. *IEEE Photonics J.* **2019**, *11*, 1–11. [CrossRef]
22. Tang, D.; Wang, X.; Zhuang, L.; Guo, P.; Yang, A.; Qiao, Y. Delay-Tap-Sampling-Based Chromatic Dispersion Estimation Method with Ultra-Low Sampling Rate for Optical Fiber Communication Systems. *IEEE Access* **2020**, *8*, 101004–101013. [CrossRef]

 *photonics*

Article

Optical Labels Enabled Optical Performance Monitoring in WDM Systems

Tao Yang [1], Kaixuan Li [1], Zhengyu Liu [1], Xue Wang [1], Sheping Shi [2], Liqian Wang [1] and Xue Chen [1,*]

1 State Key Lab of Information Photonics and Optical Communications, Beijing University of Posts and Telecommunications, Beijing 100876, China
2 ZTE Corporation, Shenzhen 518055, China
* Correspondence: xuechen@bupt.edu.cn

Abstract: Optical performance monitoring (OPM), particularly the optical power and optical signal-to-noise ratio (OSNR) of each wavelength channel, are of great importance and significance and need to be implemented to ensure stable and efficient operation/maintenance of wavelength division multiplexing (WDM) networks. However, the critical monitoring module of existing solutions generally are too expensive, operationally inconvenient and/or functionally limited to apply over WDM systems with numerous nodes. In this paper, a low-cost and high-efficiency OPM scheme based on differential phase shift keying (DPSK)-modulated digital optical labels is proposed and demonstrated. Each pilot tone is modulated by digital surveillance information and treated as an identity indicator and performance predictor that ties up to each wavelength channel and thereby can monitor the performance of all wavelength channels simultaneously by only one low-bandwidth photoelectric detector (PD) and by designed digital signal processing (DSP) algorithms. Simulation results showed that the maximum errors of channel power monitoring and OSNR estimation were both less than 1 dB after 20-span WDM transmission. In addition, offline experiments were also carried out and further verified the feasibility of our OPM scheme. This confirms that the optical label based OPM has lower cost and higher efficiency and thereby is of great potential for mass deployment in practical WDM systems.

Keywords: optical performance monitoring; wavelength division multiplexing; channel optical power; optical signal-to-noise ratio; optical labels

Citation: Yang, T.; Li, K.; Liu, Z.; Wang, X.; Shi, S.; Wang, L.; Chen, X. Optical Labels Enabled Optical Performance Monitoring in WDM Systems. *Photonics* **2022**, *9*, 647. https://doi.org/10.3390/photonics9090647

Received: 29 July 2022
Accepted: 4 September 2022
Published: 9 September 2022

1. Introduction

With the exponential growth of data traffic due to emerging bandwidth-intensive services and internet applications such as HD video streaming, cloud computing, automatic driving, 5G and other emerging applications, the optical fiber communication capacity is gradually approaching the Shannon limit [1]. Optical transport endowed with higher bit rates and better network resource utilization is an inevitable step to comply with the demand of ever-increasing capacity. Accordingly, large-capacity and long-haul WDM systems equipped with reconfigurable optical add-drop multiplexers (ROADMs) based on wavelength selective switches (WSSs) have been widely deployed in metro and backbone optical networks. The application of ROADMs could improve the flexibility and efficiency of the WDM system significantly, but the optical network would become larger and larger in scale, more and more complex in topology and dynamic in connection at the same time [2]. Therefore, to ensure its working stability and operation efficiency, low-cost and highly reliable multi-channel OPM technology is indispensable, and therefore, it is attracting a lot of attention. Especially the optical power and OSNR of each WDM channel, as key channel performance indicators, are the most important and meaningful parameters that need to be accurately monitored/estimated [3,4]. In addition, wavelength connection monitoring enables network operators to quickly discover incorrect connections caused by manual and/or software configuration errors, avoiding manual searching and recognizing, which is often time-consuming and labored. In addition, if the OPM can monitor the node location (hereinafter called "node ID") where the service signal is initially added and obtain specific surveillance information (such as symbol rate, modulation format, route configuration, etc., hereinafter called "wavelength ID"), it will be beneficial to the efficient management and optimization of WDM optical networks. However, the majority of critical monitoring modules of existing solutions, such as the conventional method using optical spectral analysis by spectrometer or optical filter, are generally too expensive, operationally inconvenient, have poor timeliness, and their functionally is limited to applied in performance monitoring of all channels over a WDM system with numerous nodes [5].

The existing mainstream monitoring schemes mainly take advantage of direct-or coherent detection of high-speed service signal and use DSP algorithms to monitor the performance of WDM channels. The monitoring technology based on direct detection without optical filters usually works at the receiver end, which divides a small part (less than 5%) of the service's optical signal for direct detection, and it analyzes the characteristics of the received signal by DSP algorithms to obtain the channel performance in a low-cost way [6,7]. However, most of the existing monitoring schemes based on direct detection inherently lose more information about the channel characteristics, and the accuracy of the monitoring is greatly reduced while the correctness of the wavelength connection also cannot be realized. On the other hand, optical filter-based monitoring technology can coherently acquire spectral information in the channel or a wavelength range through a tunable filter or diffraction grating spectrometer. In addition, due to the natural wavelength selectivity of coherent detection, the analysis of spectral information, optical power, modulation format, OSNR and other monitoring information has high recognition resolution and monitoring dynamic range [8]. However, it requires expensive narrow linewidth tunable lasers, and thereby it is difficult to apply in WDM networks due to its high cost, complex structure and one-by-one channel monitoring restrictions. Moreover, most of the existing monitoring solutions based on machine learning (ML) need to preprocess the high-speed sampling sequence with the help of DSP at the receiving end and then perform data analysis, processing and feature extraction [9,10]. The ML methods are not only difficult to effectively monitor the correctness of wavelength connections but also have higher implementation complexity than conventional DSP methods and require a large amount of data set training, parameter tuning and model generalization in advance [11]. Consequently, for a WDM network with large scale, complex structure and continuous network topology expansion, the OPM technology still presents a prominent contradiction between monitoring capability and implementation cost.

To solve these problems, OPM schemes based on optical labels have been proposed and investigated [12–20]. The typical monitoring technology based on optical label mainly adopts a pilot tone modulation to generate a low-speed label signal corresponding to the target channel, where the pilot-tone-carried label signal is loaded on the service optical signal as optical labels. At monitoring nodes, a low bandwidth direct-detection photodetector (PD) followed by the analog-to-digital converter (ADC) acquires the label signal of all wavelength channels simultaneously. The bandwidth of PD is usually at hundred-MHz order while the sampling rate of ADC is about hundred-MSa/s that is enough to demodulate the labels with relatively low cost. Theoretically, the optical labels are tied up to their corresponding optical channels anywhere in the network, and the node ID as well as wavelength ID can be flexibly loaded as optical labels; it can not only monitor the wavelength channel performance but also can reliably deliver any interested surveillance information of each channel to further sense the working state and optimize resource utilization of WDM networks.

However, when oriented to practical applications, there are still many problems in the existing solutions. On the one hand, the OPM techniques based on only pure pilot tones in [14–16] have a trade-off problem of two negative effects of stimulated Raman scattering (SRS) and chromatic dispersion. When the low frequency pilot tone is used, it will have severe SRS effects that need to be combated by dividing the whole wavelength channels into several sub-channels with the assistance of expensive optical filter. In contrast, when high-frequency pilot tones are used, the large CD caused by long-distance fiber transmission will cause severe power fading of the pilot tones, thereby significantly degrading the monitoring performance. On the other hand, the optical labelling scheme based on subcarrier index modulation (SIM) reported in [17–20] has the advantage of efficient demodulation of optical labelling signals. Nevertheless, since it adopts the SIM scheme for optical labels, not only the modulation and demodulation require large-size (16,384 points in [18]) inverse fast Fourier transform (IFFT) and fast Fourier transform (FFT), respectively, which results in very high circuit realization complexity, but also it is difficult to monitor channel optical power due to the SRS effect, and likewise, it is hard to flexibly adjust the transmission rate of label signals. In our previous work, we used pilot tones of a few MHz and demonstrated a scheme that could suppress the SRS effect on optical power monitoring based on our proposed SRS mitigation algorithm [21]. However, a scheme using such low frequency (several MHz) pilot tone has a serious SRS effect, and the monitoring performance is still difficult to guarantee even using the SRS mitigation algorithms [22,23].

In this paper, we present our recent research activities and progress on optical-label enabled OPM, and a low cost and high efficiency multi-channel optical power monitoring and OSNR estimation scheme based on differential phase shift keying (DPSK) digital optical labels is proposed and demonstrated. The proposed scheme innovatively adopts the service transmitter to flexibly load the time-domain digital label signal modulated on unique pilot tone that arranged for the corresponding wavelength channel, without additional digital to analog converter (DAC) or customized optical modulator, and without complex IFFT operation. The pilot tones have a frequency of tens of MHz that is modulated by low-speed (~K Baud) digital label information, i.e., the node ID, wavelength ID and so no. The channel optical power is monitored using the method of spectral integration, and the OSNR is estimated by calculating the amplified spontaneous emission (ASE) noise accumulation that contributed by all amplifiers. To verify the proposed scheme, we have performed simulations on WDM systems within 20 spans of fiber transmission, and the results show that the maximum error of power monitoring is less than 0.9 dB while the error of OSNR estimation is also less than 1 dB. Therefore, compared with the conventional optical filter-based OPM technology, the implementation of the proposed scheme is of lower cost and higher efficiency, which has great potential for large-scale deployment in practical WDM networks.

2. Principle of the Optical-Label Enabled OPM Scheme

The optical label is known as a top modulation or a small amount of amplitude modulation, also called low-frequency perturbation or pilot tones in some studies. The schematic diagram of the scheme is shown in Figure 1. After the digital surveillance information of node ID and wavelength ID of each wavelength channel is mapped by DPSK coding, it is modulated (~K Baud rate) to the pilot tone with a specific frequency to be the final optical label. It is noteworthy that the pilot tone frequency is larger than 40 MHz to avoid the impact of SRS crosstalk [24]. Thus, the power monitoring error caused by the SRS, resulting in the non-correspondence between the label power and the channel power, can be reasonably ignored.

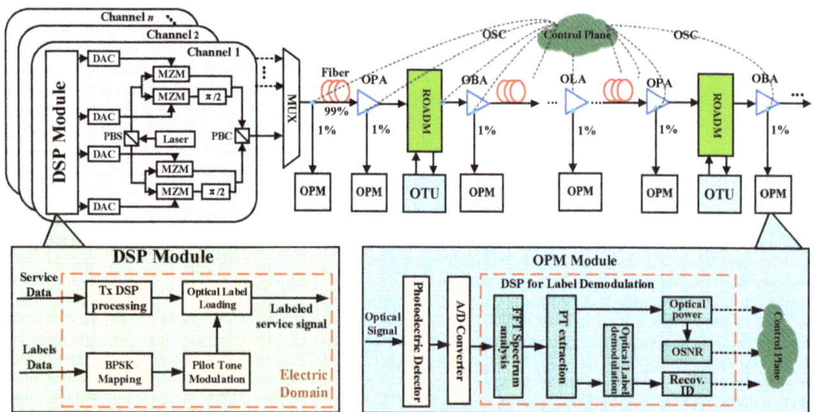

Figure 1. Schematic diagram of the optical-label enabled OPM scheme. (OTU: optical transponder unit, OPA: optical pre-amplifier, OBA: optical boost amplifier, OLA: optical line amplifier, MZM: Mach-Zehnder modulator, PBC: polarization beam combiner, PT: pilot tone, OSC: optical supervising channel).

2.1. Optical Labels Loading and Detection

At the transmitter side in Figure 1, wavelength channel 1 is taken as an example to introduce the process of label generation and modulation. In the transmitter-side DSP module, the label data are first mapped into BPSK symbols and then modulated onto pilot tones with a specific frequency to generate label signals. The label signals are subsequently loaded into the high-speed service signal in the electrical domain and thus generate a labeled service signal. It is then converted to analog signal by the DAC and drives the optical I/Q modulator to generate a labeled optical signal. Here, the DAC is the one used by the high-speed service itself, not an additional one. When arriving at optical pre-amplifier (OPA) or optical boost amplifier (OBA) in a service node, or optical line amplifier (OLA) in relay node, the received signal is divided into two parts. Thus, 99% power of the signal is used for the coherent detection of service optical signal, and 1% power of signal is used for performance monitoring and optical label demodulation. The 1% power of signal is sent into a PD to achieve optical-to-electrical conversion. The electrical signal is then converted to a digital signal by an analog-digital converter (ADC). Then, after monitoring the optical power and OSNR estimation, the node ID and wavelength ID are recovered. Finally, the monitoring results of channel optical power, OSNR and node/wavelength ID are transmitted to the control plane through optical supervising channel (OSC), and the control plane thus could not only check the route, symbol rate, modulation format and so on of each wavelength channel, but also analyze the performance of the whole WDM optical network comprehensively.

In a typical WDM system, one fiber carries the high-speed service signals of 80 wavelength channels, so different frequency PT with digital labels are modulated on corresponding wavelength channels where the same label signals are modulated on two polarizations to improve the received signal-to-noise ratio (SNR) of optical labels. For instance, the labels with PT frequency of f_1 are assigned to the service optical signal on wavelength channel 1 of λ_1, and the rest can be done in the same manner so that the digital band-pass filter in the OPM module could extract corresponding digital labels on the PT with different frequencies. The optical field of the dual polarization optical signals with amplitude-modulated optical labels can be expressed as follows:

$$E_{in,x,i} = A_{x,i}(t)\cdot \exp\{j(2\pi f_{o,i}t + \theta_{x,i}(t))\}\cdot\sqrt{[1 + m\cdot\cos(2\pi f_i t)\cdot D_i(t)]} \quad (1)$$

$$E_{in,y,i} = A_{y,i}(t)\cdot \exp\{j(2\pi f_{o,i}t + \theta_{y,i}(t))\}\cdot\sqrt{[1 + m\cdot \cos(2\pi f_i t)\cdot D_i(t)]} \quad (2)$$

where $E_{in,\frac{x}{y},i}$ represents the electric field of x/y polarization on the i-th channel, x and y represent two polarizations, n represents the serial number of the wavelength channel where it is located. $A_{x,i}(t)\cdot \exp\{j(2\pi f_{o,i}t + \theta_{x,i}(t))\}$ represents the service date signal, where $A_{x,i}(t)$ is the amplitude, $f_{o,i}$ is the center wavelength of i-th channel, $\theta_{x,i}(t)$ is the phase of service signals. $\cos(2\pi f_i t)\cdot D_i(t)$ is the optical label, where m represents the modulation depth of optical labels and f_i is the frequency of the corresponding pilot tone. It is noteworthy that the square root form of the optical label modulation on the service signal is used to facilitate analysis of label reception during square-law detection using the PD module. When it comes to the detection of optical labels in an OPM module, 1% of the optical power component is isolated and inputted into the OPM module. Direct-detection low bandwidth PD and low sample-rate ADC are used to realize the photoelectric conversion and digital sampling of the optical labels of all wavelength channels simultaneously. The PD detection can be expressed with the following equations.

$$I = \sum_{i=1}^{n}\left[|E_{in,x,i}|^2 + |E_{in,y,i}|^2\right] = \sum_{i=1}^{n}\left\{\left[A_{x,i}{}^2(t) + A_{y,i}{}^2(t)\right]\cdot[1 + m\cdot \cos(2\pi f_i t)\cdot D_i(t)]\right\} \quad (3)$$

where I represents the signal after PD square-law detection. The signal contains all the optical labels of each channel. n represents the number of PT. Here, the spectrum of optical labels is obtained and analyzed by the FFT algorithm, which is an important step for optical labels-based power monitoring. Afterward, since the digital labels are loaded by different frequency PT on different wavelength channels, digital band-pass digital filters are used here to separate the labels on different channels. Finally, differential demodulation methods are used to realize demodulation of the DPSK-based optical labels and recover the ID information. The filtered optical labels (digital labels on PT) of the i-th channel can be expressed as

$$I_i = [A_{x,i}{}^2(t) + A_{y,i}{}^2(t)]\cdot m\cos(2\pi f_i t)\cdot D_i(t) = A_i(t)\cdot m\cos(2\pi f_i t)\cdot D_i(t) \quad (4)$$

where $A_i(t)$ equals $[A_{x,i}{}^2(t) + A_{y,i}{}^2(t)]$, representing the optical label amplitude of the i-th channel after PD reception.

2.2. Channel Optical Power Monitoring

It is obvious that there is a linear proportional relationship between the channel's optical power of the service signal and the detected power of the optical labels. In addition, the integral value of the received spectrum of optical labels for a wavelength channel could represent the detected power of optical labels on the channel based on pre-tested conversion coefficients. These relationships can be expressed using Equation (5).

$$P_{ch,i} = \frac{P_{Label,i}}{m\cdot\alpha}\cdot f(V_{PD}) \quad (5)$$

where $P_{ch,i}$ and $P_{Label,i}$ represent the actual optical power of the i-th channel and the corresponding monitored optical label power respectively. m represents the modulation depth of the optical label. $f(V_{PD})$ represents the calibration coefficient of the PD module, and it depends on the photoelectric conversion characteristics of the PD itself and the feature of the optical labels. α represents the splitting ratio of the optical coupler (OC). Typically, a 1:99 (α = 1%) OC is used in our scheme to divide the labeled optical signal into OPM module.

After photoelectric conversion and digital sampling by a PD and ADC, a DSP unit is used to calculate the channel optical power. Firstly, FFT is used to obtain the spectrum of optical label, and the reception spectrum of an 80-channel WDM using 0.1 MSa/s digital labels is shown in Figure 2. It is shown that the PT frequency range of 40–55.8 MHz with 0.2 MHz grid between neighbor channels is adopted. Then, spectrum integration of each optical label spectrum from the beginning PT frequency to the ending PT frequency is carried out, respectively. For example, the channel-1 power is equal to the spectrum integration value from [39.9–40.1] MHz range. Next, a proportional coefficient of photoelectric conversion of PD is used to calculate the final channel optical power of the service signal. Finally, the monitoring results of all network monitoring nodes are transmitted to the control plane through optical supervising channel (OSC), and the control plane analyzes the performance of the whole optical network and optimizes resource utilization comprehensively.

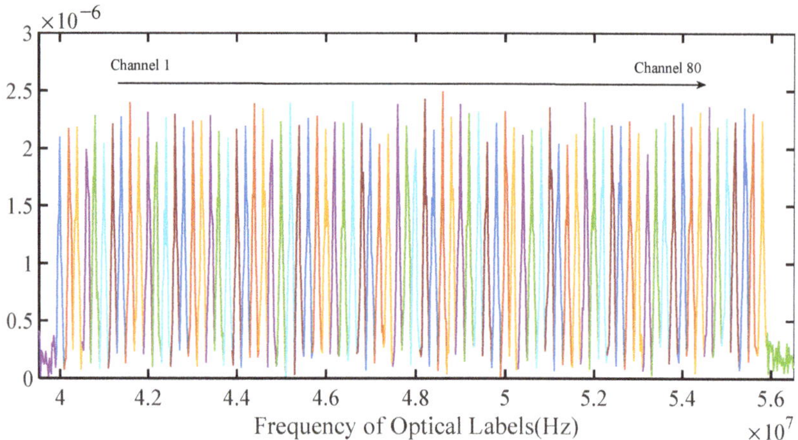

Figure 2. Spectrum of the detected optical label of 80-channel WDM systems.

2.3. OSNR Estimation

The scheme diagram of OSNR estimation is also shown in OPM Module in Figure 1. In a multi-span WDM transmission system, OSNR will gradually deteriorate due to the accumulation of ASE noise generated by the EDFA. Therefore, the OSNR could be estimated by calculating the ASE noise accumulation. In a practical, optical transmission system, the actual gain of the amplifier may be different from the configured gain, so we can calculate the actual current working gain of each optical amplifier with the help of the optical power monitored by the optical labels. At the same time, because the noise figure of optical amplifier changes with different wavelengths, here we can obtain the value of the noise figure (NF) pattern by measuring it in advance. Finally, according to the EDFA amplification of each wavelength channel, the overall amplification case of each wavelength can be obtained. Thus, based on the EDFA amplification and ASE noise accumulation of each wavelength channel, the OSNR can be estimated by the 58-Formula.

The detailed principle of OSNR estimation is as follows. For each optical amplifier in the network, its ASE noise could be expressed as [25]:

$$\mathrm{ASE(dBm)} = 10\cdot \log[(hv)(B_o)] + NF(\mathrm{dB}) + G(\mathrm{dB}) = -58(\mathrm{dBm}) + NF(\mathrm{dB}) + G(\mathrm{dB}) \quad (6)$$

where hv represents the energy of a single photon, B_o represents the reference bandwidth of OSNR which is 12.5 GHz. The NF represents the noise figure of optical amplifier. G represents the actual gain of optical amplifier which is calculated using the results of optical power monitoring. When the service signal is transmitted through a series of amplifiers, the ASE noise will be added linearly. At this time, on the service transmission link, the ASE noise accumulation by the i-th amplifier in j-th channel can be expressed as:

$$\mathrm{ASE}_{sum\_n,j}(\mathrm{dBm}) = 10\cdot \log \sum_{i=1}^{n} 10^{\frac{-58(\mathrm{dBm}) + NF_{n,i}(\mathrm{dB}) + P_{out_{n,i}}(\mathrm{dBm}) - P_{in_{n,i}}(\mathrm{dBm})}{10}} \quad (7)$$

where the $P_{out_{n,i}}$ a and $P_{in_{n,i}}$ represent the output-power and input-power of n-th amplifier in i-th channel, respectively. Finally, the OSNR of each channel can be estimated by Equation (8).

$$OSNR_{n,i}(\text{dB}) = P_{out_{n,i}}(\text{dBm}) - ASE_{sum\_n,i}(\text{dBm}) \tag{8}$$

2.4. CAPEX Analysis

Capital expenditure (CAPEX) is an important issue in optical network monitoring. How to monitor optical network with low cost and high efficiency has become the goal pursued by mainstream manufacturers and operators. We conduct a detailed cost analysis of the proposed scheme and two typical existing schemes. We compare the CAPEX of OPM schemes based on optical filter (OF) (hereinafter referred to as scheme A) [15], SIM labels (scheme B) [19] and the proposed method (scheme C). We learn the market price from a number of related device/module manufacturers and customers, and the comparison of CAPEX is mainly based on the market price of all devices/modules used in different OPM schemes. It should be noted that because the labeling rate is low, the loading process is simple, and the hardware resources required by the DSP unit are relatively small, so their cost can be reasonably ignored. Here, the cost of OPM over an 80-channels WDM system is taken as an example to compare the CAPEX. The reference cost units of different devices used in the three schemes are shown in Table 1.

Table 1. The cost of different modules used in the three OPM schemes.

Scheme A		Scheme B		Proposed C	
Modules	**Cost Units**	**Modules**	**Cost Units**	**Modules**	**Cost Units**
VOA	400 (×80)	VOA	400 (×80)	/	/
OF	9000	/	/	/	/
PD	1100	PD	1100	PD	1100
ADC	500	ADC	500	ADC	500
Total 42,600		Total 33,600		Total 1600	

Obviously, the proposed OPM scheme adopts only the service transmitter to flexibly load the time-domain digital label signals and uses a PD to detect the optical labels of all channels simultaneously. It is shown in Figure 3 that, compared with scheme A, the proposed scheme greatly reduces the CAPEX by 96% because the variable optical attenuator (VOA) and OF are completely removed. Compared with scheme B, the complex IFFT operation and VOAs are both removed in the proposed scheme, and thereby, the CAPEX is further reduced by 75%. Therefore, it could be concluded that the proposed scheme has the advantages of very low cost and high efficiency, and therefore, it has great potential in WDM network monitoring.

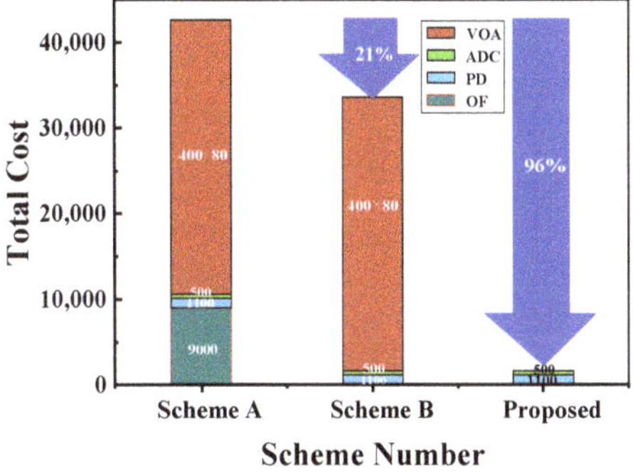

Figure 3. CAPEX comparison of three schemes.

3. Simulation and Discussion

With the help of VPI TransmissionMaker 9.0 created by VPIphotonics (Berlin, Germany) and MATLAB R2020a created by MathWorks (Natick, MA, USA), we have constructed a simulation platform of WDM transmission with OPM module based on DPSK-modulated digital optical labels to investigate the performance of the proposed scheme for channel power monitoring and OSNR estimation. Here we simulate 8, 32 and 64 channels within 20 spans WDM transmission systems. The 2 Mbps rate digital labels modulated by different PTs with 4 MHz grid are arranged to 16 G Baud polarization multiplexing (PM) QPSK and PM-16QAM transmission systems. The center wavelength of service signal is form 193.1–196.25 THz with 50 GHz interval. A photodetector with 300 MHz −3 dB bandwidth and ADC with 600 MSa/s sample rate and 10-bit resolution are used to detect the optical labels in a WDM system. Table 2 shows important simulation parameters. Here we firstly investigate the performance when using different label modulate depth of 0.05, 0.10 and 0.15. The result of optical label demodulation, such as signal noise ratio (SNR) and label power (result of spectrum integral) in QPSK and 16QAM systems, is shown in Table 3.

Table 2. Simulation parameters.

Service Modulation Format	PM-QPSK/16QAM	Modulation Depth	5%/10%/15%
Service Baud Rate	16 G Baud	EDFA Noise Figure	5 dB
Launch Power	0 dBm	Length of Each Span	100 km
Digital Label Format	DPSK	PMD Coefficient	0.1 ps/(km)$^{1/2}$
Digital Label Rate	2 Mbps	PD Bandwidth	300 MHz
Bit Number of Digital Labels	32 bit	ADC Sample Rate	600 MSa/s
Carrier Wavelength	193.1–196.25 THz	ADC Resolution	10 bit
Channel Spacing	50 GHz	Fiber Attenuation	0.2 dB/km

Table 3. SNR and power of optical labels in with different modulate depth.

Service Signal Format	Modulation Depth	SNR of Optical Label (dB)	Label Power (mW)
QPSK	5%	10.59	0.96
	10%	13.71	1.99
	15%	15.10	2.98
16QAM	5%	6.95	0.91
	10%	10.29	1.88

According to Table 3, we find that the optical label performance in QPSK system is better than that in 16QAM system, and the optical label power basically exhibits a linear relationship as the modulation depth increases. Optical label power and modulation depth m are basically linear, and the label information could be recovered accurately. In addition, the SNR of optical label under different m is sufficient for error-free demodulation.

3.1. Performence of Channel Optical Power Monitoring

In order to alleviate the influence of the photoelectric conversion characteristics of different PDs and different modulation depths on the accuracy of channel optical power monitoring, we first simulated the curve relationship of the calibration coefficient with the modulation depth in the QPSK/16QAM system. The modulation depth of the QPSK/16QAM system is increased from 2% to 15%. Then, the relationship between the calibration coefficient and the modulation depth is shown in Figure 4.

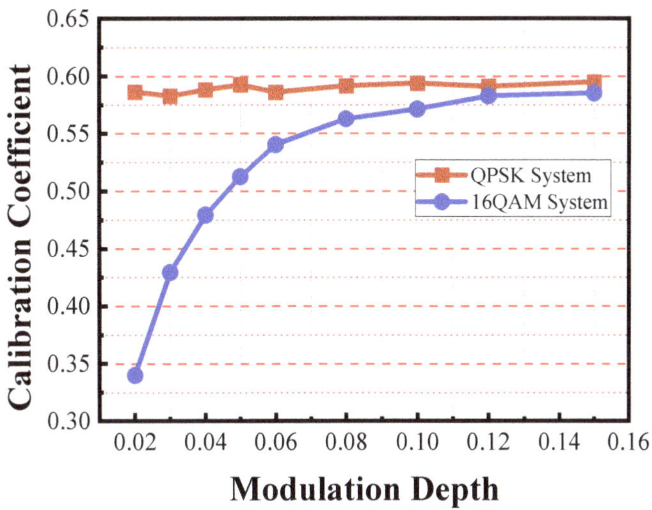

Figure 4. The calibration coefficient versus modulation depth.

Figure 3 shows that as the modulation depth increases, the calibration coefficient of the QPSK system is relatively constant with little fluctuation, but it increases first and then gradually becomes constant as the modulation depth increases in the 16QAM system. The main reasons are as follows: For the constant-module QPSK signals, the beat noise generated by PD detection is relatively small, which means that even at a low modulation depth, the noise only accounts for a small part of the entire optical label power and causes little effect on the optical label power. As the modulation depth increases, the optical label power also increases almost linearly. Therefore, the change of modulation depth hardly brings about the change of calibration coefficient.

However, for the 16QAM system with non-constant modulus, larger beat noise will be generated after PD detection. This noise accounts for a ratio of the entire optical label power at low modulation depths, even exceeding that of the digital label signal modulated by the pilot tone. This results in the optical label power being greater than the expected power at the corresponding modulation depth. In the case of monitoring the optical power of the same channel, the calibration factor is small compared with QPSK systems. The PD beat noise is almost constant or slightly increases with increasing modulation depth, but the power value of the pilot-modulated useful digital label increases linearly, resulting in a reduced effect of PD beat noise accompanied by an increase in calibration coefficients. This process gradually flattens out as the power of the digital label signal increases. When the modulation depth is increased to 15%, the effect of the beat noise is small, reaching a situation similar to that of the QPSK system, so the calibration coefficients of the two systems are almost the same at this time.

The maximum absolute errors of optical power monitoring at different modulation depths on 16QAM and QPSK systems after 20 spans fiber transmission are shown in Figure 5. It shows that the scheme of optical power monitoring using spectrum integration is feasible, and the maximum channel power monitoring error is less than 0.6 dB. The QPSK and the 16QAM systems exhibit same performance, and the maximum monitoring error is similar at different label modulation depths. Therefore, the monitoring accuracy of two systems can be considered to be roughly the same.

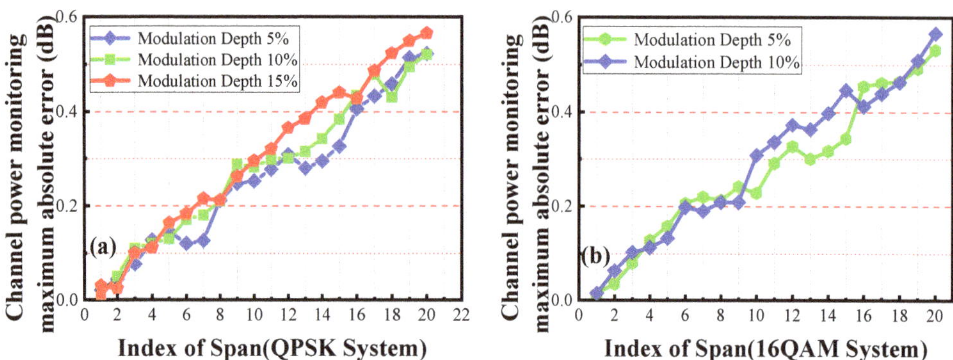

Figure 5. The optical power monitoring results of (**a**) 8-channel QPSK, (**b**) 8-channel 16QAM system.

In addition, taking a QPSK system as an example, the impact of channel number on the accuracy of channel power monitoring is also investigated. Here, the optical label modulation depth is fixed to 0.10, while the number of monitoring channels is set from 5 to 64. The monitoring error of channel optical power of different wavelength channel numbers after 20-span transmission is shown in Figure 6.

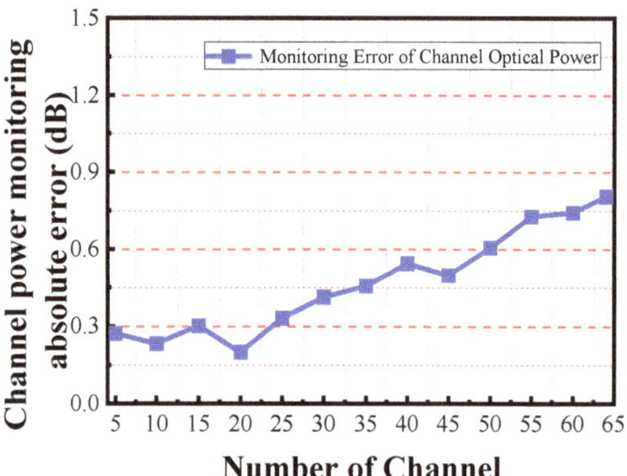

Figure 6. The power monitoring error with different channel numbers.

It can be seen form Figure 6 that when the number of channels is increased from 5 to 64, the monitoring error also increases gradually. The main reason is that as the number of wavelength channels increases, the launch power increases accordingly, and the fiber Kerr nonlinear effect increases, which deteriorates the signal quality of the optical label. In addition, as the number of wavelength channels increases, the spectrum range increases; thus, the inter-channel crosstalk caused by the SRS effect is more serious, and consequently, it will also lead to deterioration of accuracy of optical power monitoring. Moreover, we can see that the monitoring errors fluctuate in a relatively small range. This is because in each simulation with different channel numbers, the optical label sequence is randomly generated, and the random change of the label sequence characteristics causes the monitoring results to fluctuate slightly. In general, although there is a rising trend in the error of channel power monitoring, the maximum monitoring error after 20-span transmission does not exceed 0.9 dB, so it is believed that our proposed scheme is applicable to WDM optical networks.

3.2. Performence of OSNR Estimation

In order to validate the performance of OSNR estimation practically, the NF value of EDFA in each optical channel is set to 5 dB, typically. Meanwhile, the length of each span is identically set to 100 km in 20-span transmission. The input power and output power of EDFAs are monitored based on optical labels, and the OSNR in each channel is calculated by using the proposed method. Then, 16 G Baud PM-16QAM and PM-QPSK 8, 32 and 64 channels WDM transmission simulations are carried out. The result of OSNR estimation as well as the estimate error after 20-span transmission are shown in Figure 7.

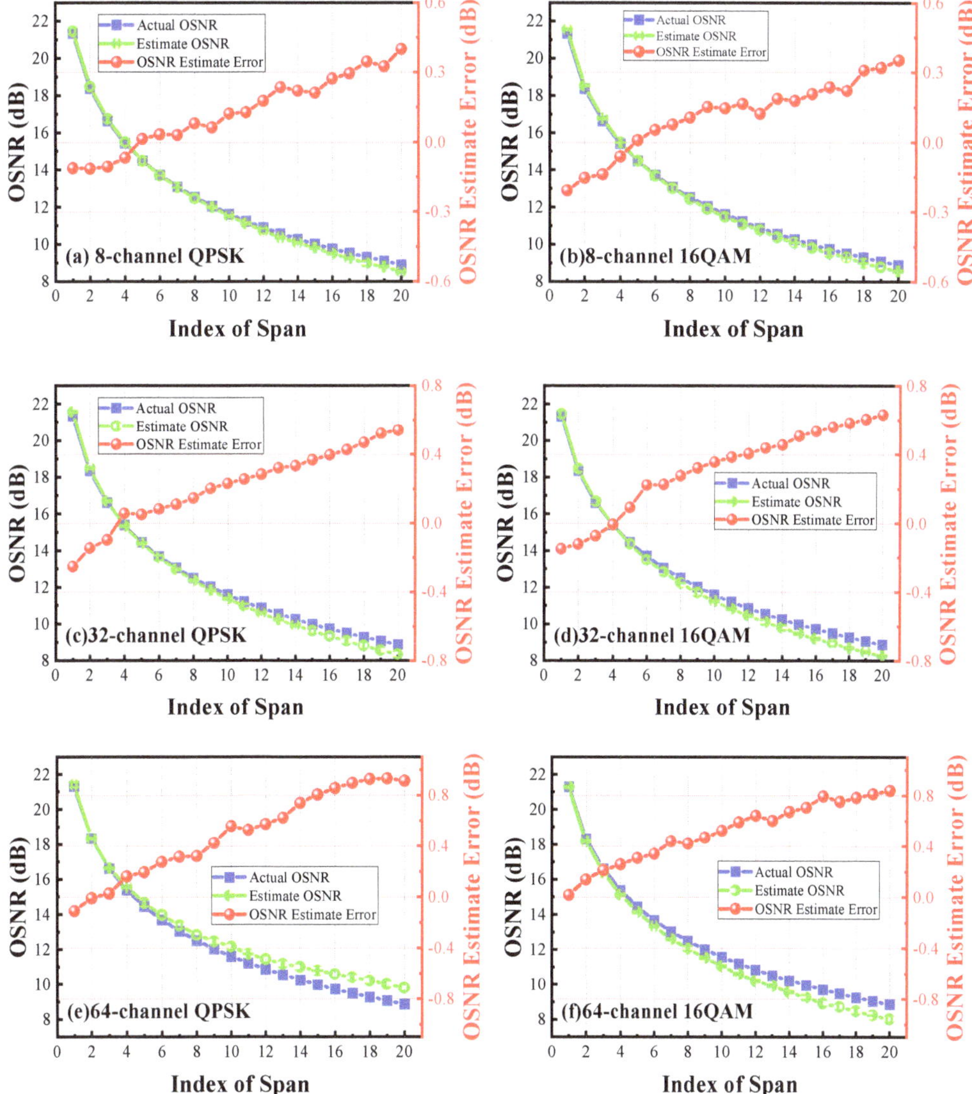

Figure 7. Result of OSNR estimation and corresponding error under 100 km equal-length span in (**a**) 8-channel QPSK, (**b**) 8-channel 16QAM, (**c**) 32-channel QPSK, (**d**) 32-channel 16QAM systems, (**e**) 64-channel QPSK systems, and (**f**) 64-channel 16QAM system.

Figure 7 shows that the scheme of OSNR estimation is feasible because that OSNR could be estimated accurately in each span both in 8/32/64-channels QPSK and 16QAM systems. It shows that the OSNR monitoring error after 20 span transmissions is less than 0.45 dB in both 8-channel PM-QPSK/16QAM systems, while it is about 0.7 dB in 32-channel PM-QPSK/16QAM systems. When the number of transmission channels increases to 64, the OSNR monitoring error after 20 span transmissions is less than 1 dB both in 64-channel PM-QPSK/16QAM systems. The main monitoring error occurs because of the error that occurred during the previous channel power monitoring. With the increase of transmission spans, the ASE noise induced by EDFA continues to accumulate due to multiple cascade amplification in the transmission link, resulting in an increase of channel power monitoring and thereby causing larger OSNR estimation error. In addition, the nonlinear effects and inter-channel cross talk also become more serious as the number of channels increases, which leads to a larger error in OSNR monitoring. Furthermore, other factors such as chromatic dispersion, polarization dependent impairments and so on will also affect the accuracy of power monitoring, which leads to a situation where the longer the transmission distance is, the greater the monitoring error is.

Meanwhile, to make the simulation more realistic and reliable, the span lengths of the 20-spans transmissions in the 8-channel system are set to 100, 60, 80, 40, 100, 50, 70, 20, 100, 60, 30, 70, 50, 80, 20, 100, 40, 100, 30 and 80 km, respectively. Thus, it is convincing to evaluate practically the OSNR estimation and corresponding errors of the target channel 1 (NF = 4) and channel 3 (NF = 5). Figure 8 shows that the maximum OSNR monitoring error is about 0.3 dB after 20-span transmission in both PM-QPSK/16QAM WDM systems. The scheme still works in the case of different span lengths, and the OSNR estimation error also tends to increase for the same reason. However, due to that the total transmission length here is reduced, resulting in a reduction of channel power monitoring, and it further makes the OSNR estimation error smaller than the results in Figure 7a,b.

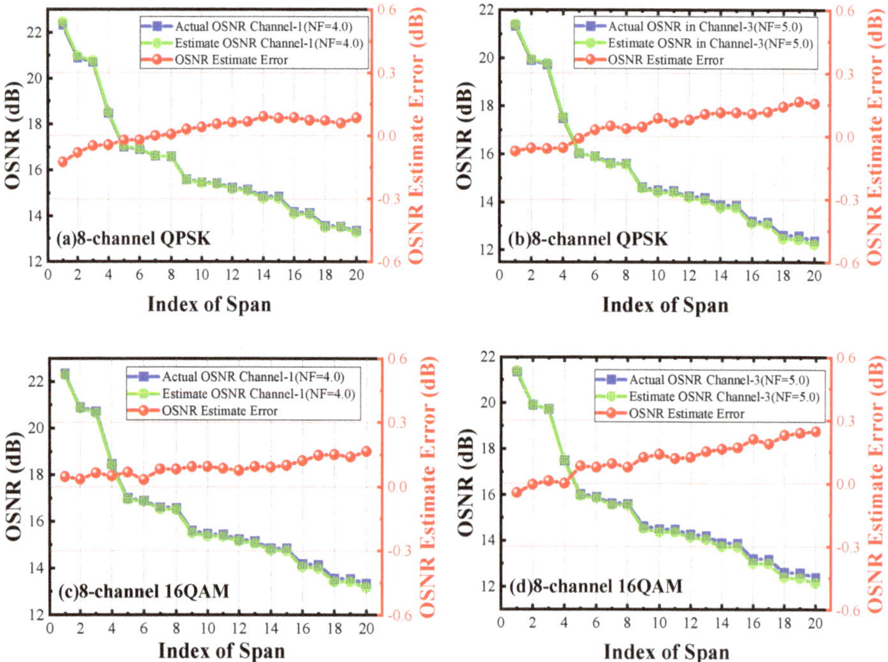

Figure 8. Result of OSNR estimation and corresponding error under [20–100 km] unequal-length span in (**a**,**b**) QPSK and (**c**,**d**) 16QAM systems.

4. Experimental Setup and Performance Analysis

To further verify the actual performance of the proposed scheme, the transmission experiment is carried out. Figure 9 shows the experimental setup of optical-label based OPM over PM-QPSK and PM-16QAM systems. Since there is only one set of laser source, optical IQ modulator, and arbitrary waveform generator (AWG) available, only single-channel optical fiber transmission and monitoring verification is carried out in this experiment. In addition, limited by the storage depth and sampling rate of the 8190A instrument (up to 12 GSa/s), the service signal rate in the offline experiment is set to 4 G Baud, and the modulation depth of the optical tag to the service signal is set to 0.10. Firstly, random bit sequences of service signal are generated and mapped into PM-QPSK/16QAM symbols. Then, we modulate the 2 Mbit/s ID information [1, 1, 1, 0, 0, 0, 1, 1, 1, 0, 0, 0, 1, 1, 1] onto a 40 MHz pilot tone and load the pilot-tone-carried labels onto the service signals by offline DSP. Next, the discrete signals after the matched filter are sent into AWG (M8190A) working at 400 MSa/s. After 80 km standard single-mode fiber (SSMF) transmission and EDFA amplification, an adjustable optical attenuator is used to change the optical power of the channel within the PD's acceptable working range of [−20–0 dBm]. A spectrometer (AQ6370) is used to observe the optical signal spectrum, while a 200 MHz-bandwidth PD is used to detect the optical labels. Then the detected 16QAM and QPSK signals are sampled by a real-time oscilloscope with 400 MSa/s for offline processing.

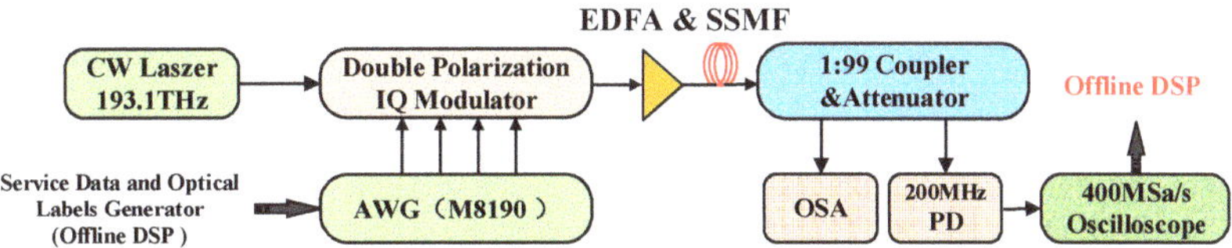

Figure 9. Diagram of experimental platform.

The waveform of the optical label after differential demodulation in offline DSP is shown in Figure 10. It shows that the transmitted DPSK digital optical labels, i.e., ID information of a wavelength channel, could be accurately recovered without error by using a low-bandwidth PD and low-sample-rate ADC with the help of DSP processing. Furthermore, we can see that the signal quality of labels in QPSK system is better than that of 16QAM. This is due to the fact that after PD reception, the amount of beat noise of QPSK constant-module signals is smaller than that of 16QAM, thereby making the QPSK system a better optical label signal-to-noise ratio when demodulating the labels.

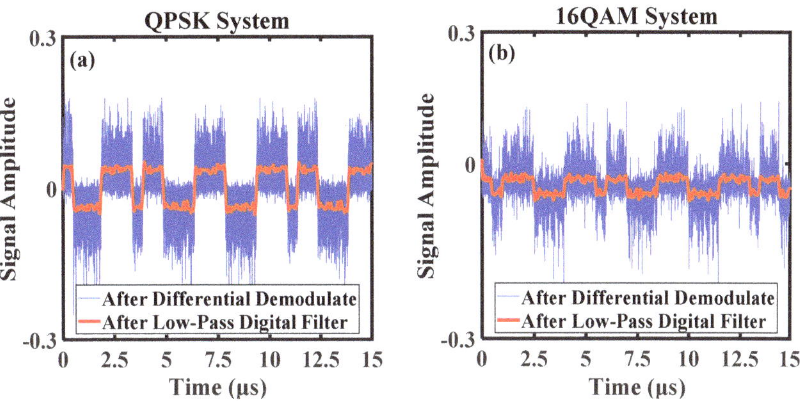

Figure 10. The demodulated waveform of optical labels in (**a**) QPSK and (**b**) 16QAM systems.

The result of power monitoring in experiments is shown in Figure 11. The result shows that the optical label power can be calculated more accurately, and then, the accurate channel optical power monitoring result can be obtained. The power monitoring error of the QPSK system is similar to that of the 16QAM system, which are both less than 0.3 dB under 80 km fiber transmission. It is important to point out that the multi-span transmission cannot be fully experimented with, resulting in better channel power monitoring performance than that of our simulation. Moreover, although the error of power monitoring fluctuates irregularly within a range, the accurate channel optical power monitoring performance is sufficient to make it a practical OPM solution.

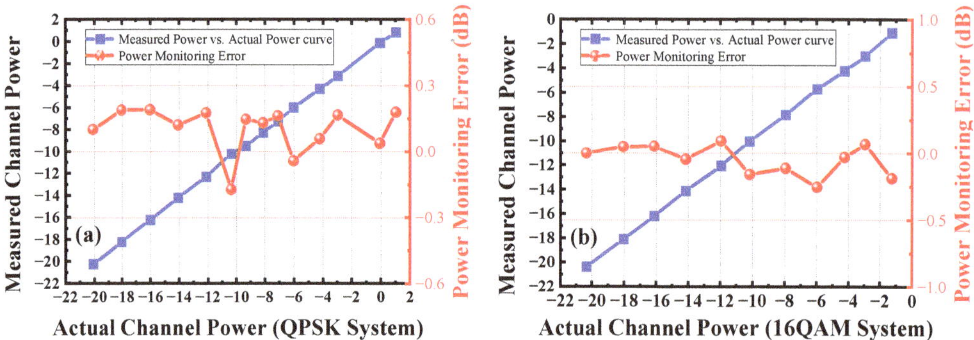

Figure 11. Results of channel power monitoring and monitoring error (**a**) in QPSK system, and (**b**) in 16QAM system.

5. Conclusions

In this paper, a low-cost and highly efficient optical labels enabled channel power monitoring and OSNR estimation of WDM system has been proposed and demonstrated. DPSK digital optical labels modulated on pilot tones with frequency of tens of MHz were applied as identity indicators and performance predictors tied up to each wavelength channel. The system can not only monitor the performance of all wavelength channels simultaneously using low-cost PD detection, but it is also able to reliably deliver any specific surveillance information to further sense the working state of WDM channels. In addition, in our proposed DSP processing in OPM modules, the channel optical power is monitored using the method of spectral integration, and the OSNR is further estimated by calculating the ASE noise accumulation form all amplifiers. The simulation results of 8, 32 and 64 channels WDM systems under 20 spans transmission show that the maximum monitoring error of channel optical power and the estimation error of OSNR are both less than 1 dB in 16 G Baud PM-QPSK and PM-16QAM systems, respectively. Furthermore, an offline experiment platform was constructed by using a PD with 300 MHz bandwidth and an ADC with 600 MSa/s sample rate, and the results show that the DPSK digital labels could be accurately recovered with the help of the proposed DSP processing, while the monitoring error of channel optical power is less than 0.3 dB. Therefore, the advantages of low cost and high efficiency make our scheme more pragmatic and more robust and therefore easier to implement and more practical for WDM system applications.

Author Contributions: Conceptualization, T.Y. and X.C.; formal analysis, T.Y., K.L. and Z.L.; investigation, T.Y. and S.S.; data curation, T.Y. and X.C.; writing—original draft preparation, T.Y., K.L. and X.W.; writing—review and editing, T.Y., L.W. and X.C.; visualization, T.Y. and K.L.; supervision, T.Y., K.L., Z.L., X.W., S.S., L.W. and X.C.; project administration, T.Y. and X.C. All authors have read and agreed to the published version of the manuscript.

Funding: This work is partly supported by National Natural Science Foundation of China (62001045), Beijing Municipal Natural Science Foundation (4214059), Fund of State Key Laboratory of IPOC (BUPT) (IPOC2021ZT17), and Fundamental Research Funds for the Central Universities (2022RC09).

Institutional Review Board Statement: Not applicable.

Informed Consent Statement: Not applicable.

Data Availability Statement: The data presented in this study are available on request from the corresponding author. The data are not publicly available due to the data also forms part of an ongoing study.

Acknowledgments: The authors express their appreciation to reviewers for their valuable suggestions.

Conflicts of Interest: The authors declare no conflict of interest.

References

1. Miladić-Tešić, S.; Marković, G.; Peraković, D.; Cvitić, I. A review of optical networking technologies supporting 5G communication infrastructure. *Wirel. Net.* **2022**, *28*, 459–467. [CrossRef]
2. Ji, Y.; Zhang, J.; Xiao, Y.; Liu, Z. 5G flexible optical transport networks with large-capacity, low-latency and high-efficiency. *China Commun.* **2019**, *16*, 19–32. [CrossRef]
3. Pan, Z.; Yu, C.; Willner, A.E. Optical performance monitoring for the next generation optical communication networks. *Opt. Fiber Technol.* **2010**, *16*, 20–45. [CrossRef]
4. Dong, Z.; Khan, F.N.; Sui, Q.; Zhong, K.; Lu, C.; Lau, A. Optical performance monitoring: A review of current and future technologies. *J. Lightwave Technol.* **2016**, *34*, 525–543. [CrossRef]
5. Dong, Y.; Jiang, T.; Teng, L.; Zhang, H.; Chen, L.; Bao, X.; Lu, Z. Sub-MHz ultrahigh-resolution optical spectrometry based on Brillouin dynamic gratings. *Opt. Lett.* **2014**, *39*, 2967–2970. [CrossRef] [PubMed]
6. Calvin, C.C.K. *Optical Performance Monitoring: Advanced Techniques for Next-Generation Photonic Networks*, 1st ed.; Academic Press: Cambridge, MA, USA; Elsevier: New York, NY, USA, 2010; ISBN 978-0-12-374950-5.
7. Cui, S.; Jin, S.; Xia, W.; Ke, C.; Liu, D. Improved symbol rate identification method for on–off keying and advanced modulation format signals based on asynchronous delayed sampling. *Opt. Commun.* **2015**, *354*, 218–224. [CrossRef]
8. Amrani, A.; Junyent, G.; Prat, J.; Comellas, J.; Ramdani, I.; Sales, V.; Roldan, J.; Rafel, A. Performance monitor for all-optical networks based on homodyne spectroscopy. *IEEE Photonics Technol. Lett.* **2000**, *12*, 1564–1566. [CrossRef]
9. Khan, F.N.; Zhong, K.; Al-Arashi, W.H.; Yu, C.; Chao, L.; Lau, A. Modulation format identification in coherent receivers using deep machine learning. *IEEE Photonics Technol. Lett.* **2016**, *28*, 1886–1889. [CrossRef]
10. Du, J.; Yang, T.; Chen, X.; Chai, J.; Shi, S. A CNN-based cost-effective modulation format identification scheme by low-bandwidth direct detecting and low rate sampling for elastic optical networks. *Opt. Commun.* **2020**, *471*, 126007. [CrossRef]
11. Mata, J.; de Miguel, I.; Durán, R.J.; Merayo, N.; Singh, S.K.; Jukan, A.; Chamania, M. Artificial intelligence (AI) methods in optical networks: A comprehensive survey. *Opt. Switch Netw.* **2018**, *28*, 43–57. [CrossRef]
12. Park, K.J.; Shin, S.K.; Chung, Y.C. Simple monitoring technique for WDM networks. *Electron. Lett.* **1999**, *35*, 415–417. [CrossRef]
13. Jun, S.B.; Kim, H.; Park, P.K.J.; Lee, H.; Chung, Y.C. Pilot-tone-based WDM monitoring technique for DPSK systems. *IEEE Photonics Technol. Lett.* **2006**, *18*, 2171–2173. [CrossRef]
14. Jiang, Z.; Tang, X. Low-cost signal spectrum monitoring enabled by multiband pilot tone techniques. In Proceedings of the 2018 European Conference on Optical Communication (ECOC), Roma, Italy, 23–27 September 2018; pp. 1–3.
15. Ji, H.C.; Park, K.J.; Lee, J.H.; Chung, H.S.; Son, E.S.; Han, K.H.; Jun, S.B.; Chung, Y. Optical performance monitoring techniques based on pilot tones for WDM network applications. *J. Opt. Netw.* **2004**, *3*, 510–533. [CrossRef]
16. Liu, Z.; Yang, T.; Shi, S.; Du, J.; Wang, J. A Highly Reliable Timing Error Tolerated Optical Label Demodulation Algorithm for WDM Optical Network Monitoring. In Proceedings of the Asia Communications and Photonics Conference 2021, Shanghai, China, 24–27 October 2021; pp. 1–3.
17. Yang, C.; Li, X.; Luo, M.; He, Z.; Li, H.; Li, C.; Yu, S. Optical Labelling and Performance Monitoring in Coherent Optical Wavelength Division Multiplexing Networks. In Proceedings of the 2020 Optical Fiber Communications Conference and Exhibition (OFC), San Diego, CA, USA, 8–12 March 2020; pp. 1–3.
18. Yang, C.; Luo, M.; Zhang, X.; Meng, L.; Luan, Y.; Mei, L.; He, Z. Demonstration of Real-time Optical Labelling System for Coherent Optical Wavelength Division Multiplexing Networks. In Proceedings of the Asia Communications and Photonics Conference (ACP) and International Conference on Information Photonics and Optical Communications (IPOC) 2020, Beijing, China, 24–27 October 2020; pp. 1–3.
19. Yang, C.; Luo, M.; He, Z.; Xiao, X. Flexible and Transparent Optical Labelling in Coherent Optical Wavelength Division Multiplexing Networks. In Proceedings of the 2022 Optical Fiber Communication Conference (OFC), San Diego, CA, USA, 6–10 March 2022; pp. 1–3.
20. Zhang, X.; Yang, C.; Luo, M.; Meng, L.; Jiang, F.; He, Z. Real Time Optical Label System for Coherent Optical Wavelength Division Multiplexing Networks. In Proceedings of the 26th Optoelectronics and Communications Conference (OECC), Hong Kong, China, 3–7 July 2021; pp. 1–3.
21. Du, J.; Yang, T.; Shi, S.; Chen, X.; Wang, J. Optical Label-enabled Low-cost DWDM Optical Network Performance Monitoring Using Novel DSP Processing. In Proceedings of the Asia Communications and Photonics Conference 2020, Beijing, China, 24–27 October 2020; pp. 1–3.

22. Park, P.K.J.; Kim, C.H.; Chung, Y.C. Performance analysis of low-frequency pilot-tone-based monitoring techniques in amplified wavelength-division-multiplexed networks. *Opt. Eng.* **2008**, *47*, 025009. [CrossRef]
23. Chung, H.S.; Shin, S.K.; Park, K.J.; Woo, H.G.; Chung, Y.C. Effects of stimulated Raman scattering on pilot-tone-based WDM supervisory technique. *IEEE Photonics Technol. Lett.* **2000**, *12*, 731–733. [CrossRef]
24. Jiang, Z.; Tang, X.; Wang, S.; Gao, G.; Jin, D.; Wang, J.; Si, M. Progresses of Pilot Tone Based Optical Performance Monitoring in Coherent Systems. *J. Lightwave Technol.* **2022**, *40*, 3128–3136. [CrossRef]
25. Keiser, G. *Optical Fiber Communications*, 4th ed.; Publishing House of Electronics Industry: Beijing, China, 2012; ISBN 978-7-121-16171-1.

MDPI

Article

Parallel Distribution Matcher Base on CCDM for Probabilistic Amplitude Shaping in Coherent Optical Fiber Communication

Yao Zhang [1], Hongxiang Wang [1,*], Yuefeng Ji [1] and Yu Zhang [2]

[1] State Key Laboratory of Information Photonics and Optical Communications, School of Information and Communication Engineering, Beijing University of Posts and Telecommunications, Beijing 100876, China
[2] Academy of Broadcasting Science, NRTA, Beijing 102206, China
* Correspondence: wanghx@bupt.edu.cn

Abstract: As a typical high-order modulation format optimization technology, constellation probability shaping enhances generalized mutual information (GMI) by optimizing the probability distribution of each constellation point of the signal. It can improve the transmission capacity of the same order M Quadrature Amplitude Modulation (QAM) signal under the condition of limited average transmission power, and further narrow the gap with the Shannon limit capacity. The distribution matcher is a key part of constellation probability shaping since it not only ensures the one-to-one mapping of input and output sequences but also realizes the function of probability shaping. The constant composition distribution matcher (CCDM) structure is a widely utilized distribution matcher in the current probability shaping technology. Based on CCDM, a parallel distribution matcher scheme is proposed in this paper. It has a lower rate loss than CCDM for short output lengths (n is less than 100). Block lengths can be reduced by up to 30% with the same rate loss. When the GMI is the same as for the probability shaping (PS) 64QAM signal using CCDM, the OSNR required by the PS-64QAM signal using this scheme can be enhanced by 0.12dB, the block length can be reduced by 40%, and the transmission distance in a standard single-mode fiber can be slightly extended.

Keywords: probabilistic amplitude shaping; constant composition distribution matcher; higher order modulation format signal; constellation shaping; LDPC coding

Citation: Zhang, Y.; Wang, H.; Ji, Y.; Zhang, Y. Parallel Distribution Matcher Base on CCDM for Probabilistic Amplitude Shaping in Coherent Optical Fiber Communication. *Photonics* **2022**, *9*, 604. https://doi.org/10.3390/photonics9090604

Received: 31 May 2022
Accepted: 23 August 2022
Published: 25 August 2022

1. Introduction

The communication transmission rate and data flow in the network continue to expand as the Internet of Everything era dawns [1,2]. Optical fiber communication network is an essential supporting platform in the process of transmitting and exchanging information. Continuously improving its system capacity is the eternal goal of the development of optical communication. The high-order modulation format optical signal carries more bit information per symbol, boosting spectral efficiency and transmission capacity, and is the primary modulation method used in the current high-speed and large-capacity coherent optical communication system. As the probability of the appearance of each symbol in the standard M-QAM modulation format optical signal is the same, the system capacity is difficult to approach the Shannon limit. The gap between the standard M-QAM signal and the Shannon capacity in the additive white Gaussian noise channel is around 1.53 dB [3]. As a typical high-order modulation format optimization technology, constellation shaping enhances mutual information and generalized mutual information by optimizing the distribution of each constellation point of the signal. It includes geometric shaping (GS) and probability shaping (PS), which can improve the transmission capacity of the same order M-QAM signal under the condition of limited average transmission power and further narrow the gap with Shannon limit capacity.

In comparison with geometric shaping, the position of each constellation point of the signal remains the same after probability shaping, but the transmission probability of different symbols varies. As a result, it is compatible with the existing modulation receiving system and digital signal processing technology, and can be flexibly combined with a range of multiplexing methods and channel coding technology, with low system complexity [3], this paper focuses on constellation probability shaping.

When the source obeys the continuous Gaussian distribution, the largest capacity can be achieved in the AWGN channel with power constraint P, according to Shannon [4]. For the probability shaping technology, the constellation points are dispersed discretely, making it difficult to determine what kind of distribution to use to approach the continuous Gaussian distribution. Kschischang demonstrated in 1993 that the Maxwell–Boltzmann (MB) distribution is the optimum for constellation probability shaping [5]. Subsequently, many probability shaping schemes, such as Gallager's Scheme [6–8], Trellis Shaping [9], Concatenated Shaping [10], and Bootstrap Scheme [11], have been proposed to realize the shaping and optimization of QAM signals. P. Schulte and G. Böcherer innovatively proposed a probabilistic amplitude shaping (PAS) architecture in 2015 that combined a constant-composition distribution-matcher (CCDM) with forward error correction (FEC) encoding to implement PS [12]. Coding and shaping are decoupled in the PAS scheme thanks to a parallel transmitter design, which substantially simplifies the implementation of encoder and decoder and makes PS techniques practical. Many researchers have proven that probability shaping technology improves transmission distance, spectrum efficiency, and bit error performance in communication systems using the PAS scheme [13–18]. In the present high-order QAM modulation system, it is one of the preferred technologies.

The distribution matcher is a key component of PAS since it not only ensures the one-to-one mapping of input and output sequences but also realizes the function of probability shaping. The CCDM structure [19] is a widely utilized distribution matcher in the current probability shaping technology. It is based on arithmetic encoding, and the rate loss of CCDM can only tend to zero as the output symbol length approaches infinity. Hardware implementation is relatively complex in today's high-speed optical communication system, and the complexity increases linearly with the length of symbols. Furthermore, arithmetic coding is a highly serial coding method. It divides the interval into intervals to reflect the input and output sequences. While the length of input and output sequences is long, the interval that must be divided grows, the borders between intervals blur, it is difficult to distinguish when mapping, and mistakes are common. It is, therefore, necessary to propose a distribution matcher with better performance even when the output symbol length is short.

A multiset-partition distribution matcher (MPDM) has been proposed to reduce the needed output lengths and can be considered as layered CCDM operations [20,21]. MPDM also adopts the idea of pairwise optimization; it needs to build a Huffman tree structure to index different complementary pairs and uses serial processing for the input data. The coding process is complex and error-prone. Many distributed matcher techniques with a parallel structure have been proposed to address the extremely serial coding approach of CCDM, such as bit-level distribution matcher [22,23], parallel-amplitude distribution matcher [24], and Hierarchical distribution matcher [25]. In addition, widely used enumerative sphere shaping (ESS)[26] and shell mapping (SM) are notable sphere shaping (SpSh) methods, which are different from CCDM. The SpSh considers amplitude sequences in a sphere. ESS orders these sequences lexicographically, while SM orders them based on their energy. Compared with ESS and SM, CCDM has similar latency and is significantly superior in terms of storage complexity [27].

In this work, we propose a parallel distribution matcher based on CCDM to improve the performance of CCDM at short block lengths. Contrary to the constant composition of the CCDM, this DM can output variable composition by using pairwise optimization and parallel structure. The proposed structure has a lower rate loss than CCDM for short output lengths (n less than 100), and the output block lengths can be lowered by up to 30% with the same rate loss. When the value of generalized mutual information is the same as it is when using CCDM, the OSNR tolerance of the PS-64QAM signal using the proposed approach can be enhanced by 0.12 dB and the block length can be reduced by 40%. At the same time, the transmission distance in the standard single-mode fiber link becomes longer.

2. Fundamentals of Probabilistic Shaping

2.1. Constant Composition Distribution Matcher

The function of constant composition distribution is to convert binary bit sequences $B^k = b_1 b_2 \ldots b_k$ into amplitude sequences $A^n = a_1 a_2 \ldots a_n$ in a one-to-one, reversible mapping. The term "constant composition" has two meanings: the length of the input bit sequence and the output amplitude sequence are both fixed, as is the composition of the output amplitude sequence, which means the probability of each amplitude occurring is constant.

Specifically, the input sequence B^k is a uniformly distributed binary sequence of length k. If we define a set of amplitudes $\chi = \{\alpha_1, \alpha_2, \ldots, \alpha_i\}$, then the output sequence A^n is of length n and the probability of each amplitude will be $P_A(\alpha_i) = \frac{n_{\alpha_i}}{n}$, indicating that the amplitude α_i will occur n_{α_i} times in this sequence with length n. Use $\Gamma^n_{P_A}$ to denote the set of all sequences of length n with a probability P_A of each amplitude. For example, for a 4ASK signal, $\chi = \{1, 3\}$; when n = 4, P(1) = 3/4, P(3) = 1/4, $\Gamma^n_{P_A} = \{1113, 1131, 1311, 3111\}$, the CCDM needs to map the amplitude sequence in $\Gamma^n_{P_A}$ with a sequence of binary bits of length k, $k = \log_2 \Gamma^n_{P_A} = 2$, such as the following mapping: $00 \Leftrightarrow 1113, 01 \Leftrightarrow 1131, 10 \Leftrightarrow 1311, 11 \Leftrightarrow 3111$.

The key steps for constructing a CCDM are as follows: specify the probability $P_A(\alpha_i)$ of the amplitude α_i and the length n of the output amplitude sequence; the number of input bits for CCDM of a set $\Gamma^n_{P_A}$ is given by (1). $\lfloor \cdot \rfloor$ is rounded down. Since CCDM is an invertible mapping, the size of the codebook C_{ccdm} should be 2^k; select a subset of $\Gamma^n_{P_A}$ as the codebook and establish the mapping relation f_{ccdm} to finish one-to-one mapping of the input and output sequences: $\{0,1\}^k \xleftrightarrow{f_{ccdm}} C_{ccdm} \subseteq \Gamma^n_{P_A}$.

$$k = \left\lfloor \log_2 \Gamma^n_{P_A} \right\rfloor = \left\lfloor \log_2 \frac{n!}{\prod_{i=1}^{i} n_{\alpha_i}!} \right\rfloor, \; n_{\alpha_i} = n \cdot P_A(\alpha_i), \tag{1}$$

Any finite length distribution matcher exists a rate loss problem. The rates of CCDM are $R = k/n$ (bit/symbol), which means that each symbol in the output amplitude sequence carries k bits of information. Under the corresponding probability distribution, the information that each symbol is theoretically able to carry is $H(A)$, with a rate loss of (2).

$$R_{loss} = H(A) - R = H(A) - \frac{k}{n}, \tag{2}$$

Similar to the above 4ASK signal, P(1) = 3/4, P(3) = 1/4, H(A) = 0.8113 bit/symbol; when n = 20, $\left\lfloor \log_2 \Gamma^n_{P_A} \right\rfloor = 13$, $R_{loss} = 0.8113 - 0.65 = 0.1613$; when n reaches 10,000, R = 8106/10,000 = 0.8106 bit/symbol. So, CCDM is an asymptotically optimal mapping scheme when the output sequence length $n \to \infty$, R $\to$ H(A). When n is relatively small, CCDM has a large rate of loss.

2.2. Probabilistic Amplitude Shaping (PAS) Scheme

The data are modulated to the two polarizations of the optical carrier in the polarization multiplexing coherent optical transmitter, and the in-phase and quadrature components of the optical carrier are modulated separately in each polarization; so, the constellation of two-dimensional square m^2-QAM signal can be expressed as the Cartesian product of the one-dimensional m-ASK signal. Use the variable X to represent the constellation point of the m-ASK signal, and use x to represent the specific value. Take x = ±1, ±3, . . . , ±(m − 1); when the gray mapping is utilized, each constellation point can be stated as the product of magnitude and sign and is independent—that is, X = A * S, where

A = 1, 3, ..., (m−1) and S = 1 or −1. Since each constellation point is symmetrical about the origin, the probabilities of the symmetrical constellation points are equal, $P_X(x) = P_X(-x)$, $P_S(1) = P_S(-1) = 1/2$, the probability shaping of the two-dimensional square m^2-QAM signal can be converted into the probability shaping of the one-dimensional m-ASK signal, which can be further converted into probabilistic shaping of the positive amplitude of the m-ASK signal, hence the name probabilistic amplitude shaping.

Figure 1a shows the principle diagram of the probability amplitude shaping using a constant composition distribution matcher and low-density parity check code. Taking the 64QAM modulation format signal as an example, the I channel or the Q channel is an 8ASK signal; the generating process of the bit sequence corresponding to the PS-64QAM signal is shown in Figure 1b.

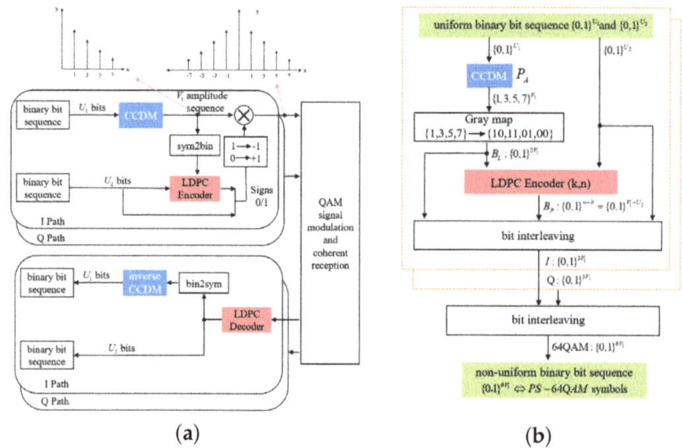

Figure 1. (**a**) The principle diagram of probability amplitude shaping using a CCDM and LDPC. (**b**) Bit sequence generation process of PS-64QAM signal.

A sequence of binary data of length U_1 at the input is fed into CCDM to generate a sequence of amplitudes of length V_1, the amplitudes are picked from $\{1,3,5,7\}$, and the number of occurrences of each amplitude are equal to the target probability. The rate of CCDM is $R_{DM} = U_1/V_1$. Then, apply the gray map on the V_1 amplitude sequences, where the mapping rule is $\{1,3,5,7\} \Leftrightarrow \{10,11,01,00\}$, the mapped binary bit sequence is B_L, and the length is $m * V_1$ (m = 2). Next, the binary information sequences of length U_2 at the input and B_L are concatenated and encoded together in LDPC to create check bits B_P of length $n - k = V_1 - U_2$, the encoding rate determined by (3). Finally, U_2 and B_p are utilized as the sign bit together and B_L is used as the amplitude bit, which are, respectively, assigned to different positions of the codeword by the interleaver. The interleaved bitstream can be used as data information for the I channel or the Q channel to modulate the QAM signal. At the receiving end, the input binary bit sequence can be recovered by performing the reverse process. In the following sections, the mutual information and bit error rate performance of PS-QAM signals will be analyzed based on this probabilistic amplitude shaping scheme.

$$R_c = k/n = (m * V_1 + U_2)/(V_1 - U_2 + k) = (m * V_1 + U_2)/(m+1) * V_1, \quad (3)$$

3. Parallel Distribution Matching Based on CCDM

When the length of the output block (the length of the output amplitude sequence) n is small, CCDM suffers from rate loss. In this section, a paired optimization method is adopted to solve this issue, and a parallel distribution matcher structure based on CCDM is proposed.

3.1. Paired Optimization Principle

When the length n of CCDM output amplitude sequence is small, the probability distribution of each amplitude must be quantized—that is, $P_A \to P_{A'}$, to ensure that the number of occurrences of each amplitude is an integer and the quantization should try to satisfy that P_A and $P_{A'}$ are as close as possible.

When the output amplitude sequence of the CCDM is $A^n = a_1 a_2 \ldots a_n$, the length is n, the amplitude setting is $\chi = \{\alpha_1, \alpha_2, \ldots, \alpha_i\}$, each amplitude value in the sequence A^n is taken from χ, the number of occurrences of the amplitude α_i is n_{α_i}, and $n_{\alpha_i} = n \cdot P_{A'}(\alpha_i)$; the set $C = \{n_{\alpha_1}, n_{\alpha_2}, \ldots, n_{\alpha_i}\}$ is called $P_{A'}$ composition with a probability distribution a, and the size of the set C can be determined using the permutation and combination formula as shown in (4), which is equivalent to $\Gamma^n_{P_A}$.

$$M(C) = \frac{n!}{n_{\alpha_1}! n_{\alpha_2}! \ldots n_{\alpha_i}!} \tag{4}$$

$C_{typ} = \{n_{\alpha_1}, n_{\alpha_2}, \ldots, n_{\alpha_i}\}$ represents a typical composition of CCDM output, and obeys the probability distribution $P_{A'(\alpha_i)}$. The pairwise optimization means that the single output amplitude sequence of the CCDM does not follow the probability distribution $P_{A'(\alpha_i)}$—that is, $C \neq C_{typ}$. However, the average composition of many output amplitude sequences is equal to C_{typ}, with all output amplitude sequences obeying the target probability $P_{A'(\alpha_i)}$ as a whole. The purpose of pairwise optimization is to find components that satisfy (5) [20], where l represents the N_{comp} possible composition of the output amplitude sequence and c_l represents the number of occurrences of the sequence whose composition is C_l in the output sequence; its value does not exceed $M(C_l)$ at most. Here, we only consider the case of pairwise complementarity, not the case of three or more compositions of complementarity.

$$\frac{\sum_{l}^{N_{comp}} c_l \cdot C_l}{\sum_{l}^{N_{comp}} c_l} = C_{typ} \tag{5}$$

For example, the length of the output amplitude sequence of CCDM is n = 10; the amplitude is taken from (a_1, a_2, a_3, a_4), assuming the probability distribution after quantization is $P_{A'} = (0.4, 0.3, 0.2, 0.1)$; $C_{typ} = (4, 3, 2, 1)$; and the information entropy H(A) = 1.85 bit/symbol. If only C_{typ} is considered in the output composition of CCDM, $M(C_{typ}) =$ 12,600, it can map $k = \lfloor \log_2(C_{typ}) \rfloor = 13$ input bits, the rate is 1.3 bit/symbol, and the rate loss is 0.55 bit. When considering the composition of $C_1 = (4, 2, 3, 1)$ and $C_2 = (4, 4, 1, 1)$, although a single output amplitude sequence is not equal to $P_{A'}$, but $M(C_1) = M(C_2) =$ 6300, the output amplitude sequences of C_1 and C_2 can be complementary, and the overall obeys the probability distribution $P_{A'}$; thus, the total output amplitude sequence is equal to 12,600 + 6300 + 6300 = 25,200, which can map 14-bit input binary sequence. Compared with only considering C_{typ}, the rate is increased by 0.1 bit/symbol [20].

3.2. Implementation of Parallel Distribution Matcher Base on CCDM

Based on the above principle, the benefit of pairwise optimization can be exploited to improve the rate penalty of CCDM. For simplicity, only two complementary components are considered here satisfying (6).

$$C_l + \overline{C_l} = 2C_{typ} \tag{6}$$

All complementary pairs can be found by exhaustive methods, but when n is relatively large, there are many complementary pairs and it is easy to miss. A summary formula is given below, which can be used to locate all complementary pairs $\{C_l, C_2\}$ that meet (6) on a regular basis. When n = 10, $C_{typ} = (4,3,2,1)$, $2C_{typ} = (8,6,4,2)$, consider the following polynomial:

$$x_1 + x_2 + x_3 + x_4 = 10\ 0 \leqslant x_1 \leqslant 8,\ 0 \leqslant x_2 \leqslant 6,\ 0 \leqslant x_3 \leqslant 4,\ 0 \leqslant x_4 \leqslant 2$$
$$\underbrace{(x^0+x^1+x^2+x^3+x^4+x^5+x^6+x^7+x^8)}_{x_1} * \underbrace{(x^0+x^1+x^2+x^3+x^4+x^5+x^6)}_{x_2} * \underbrace{(x^0+x^1+x^2+x^3+x^4)}_{x_3} * \underbrace{(x^0+x^1+x^2)}_{x_4} \quad (7)$$

For example, x^8 in x_1, x^1 in x_2, x^1 in x_3, and x^0 in x_4 is equal to $x^{8+1+1+0} = x^{10}$ when multiplied; so, the composition C_1 = (8, 1, 1, 0) is one of the cases, and the corresponding complementary pair is $\overline{C_1}$ = (0, 5, 3, 2). This polynomial multiplication enables us to find regularly all the possible compositions and the corresponding complementary pairs.

Figure 2 shows the block diagram of the parallel distribution matcher. The length of the input binary sequence is $k = p_l + k_a + k_b$, which the first p_l bits utilized to select complementary pairs, and the remaining $k_a + k_b$ bits are serial-to-parallel conversion as needed. In the upper $CCDM_1$, the bit sequence of length k_a is mapped to the 2^{k_a} amplitude sequences in the set $M(C_l)$ by arithmetic coding, and in the lower $CCDM_2$, the bit sequence of length k_b is also mapped to the 2^{k_b} amplitude sequences in the set $M(\overline{C}_l)$. After the parallel-to-serial conversion of the generated two-paths amplitude sequence, the component composition is $2C_{typ}$, which obeys the target probability distribution $P_{A'}$.

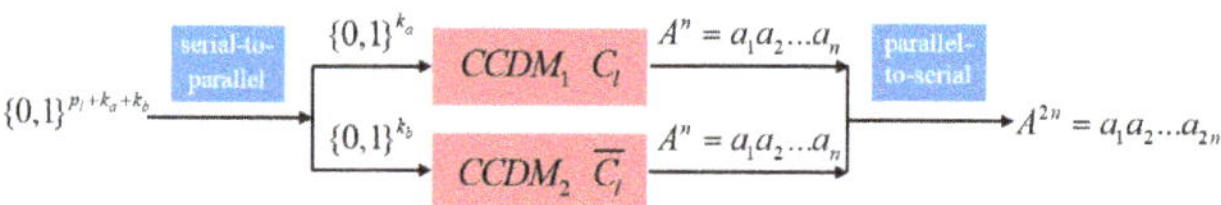

Figure 2. The block diagram of the parallel distribution matcher.

The specific implementation steps are as follows:

(1) Set the length of the output blocks of the two CCDMs to be n, quantify the target probability distribution $P_A \to P_{A'}$ to ensure that the number of occurrences of each amplitude is an integer, and obtain C_{typ} and $k_{typ} = \lfloor \log_2 M(C_{typ}) \rfloor$.

(2) According to (7), find all qualified complementary pairs C_l, $M(\overline{C}_l)$, denoted as N_{pair}; calculate $M(C_l)$ and $M(\overline{C}_l)$, and obtain the mappable input bit sequence length $k_a + k_b = \lfloor \log_2 M(C_l) \rfloor + \lfloor \log_2 M(\overline{C}_l) \rfloor$; sort N_{pair} complementary pairs from large to small according to the size of $k_a + k_b$—the complementary pairs of $k_a + k_b < k_{typ}$ are preferentially discarded because such complementary pairs will reduce the overall rate loss, and there are N'_{pair} types of complementary pairs left after discarding; to obtain $p_l = \log_2 N'_{pair}$, a binary sequence of length p_l and 2^{p_l} complementary pairs maintain a one-to-one mapping relationship.

(3) At the input end, for a string of input binary bit sequences, first determine which complementary pair is used by the two CCDMs according to the size of the first p_l bits. Then, make a serial–parallel conversion after obtaining $k_a + k_b$; the corresponding amplitude sequences are obtained through arithmetic coding, respectively. Finally, the two amplitude sequences are converted in parallel to serial, which can then be employed in the subsequent probability amplitude shaping system to achieve the probability shaping of QAM signals.

(4) The appropriate complementary pairs are determined by counting the frequency of occurrence of each amplitude in n amplitude sequences while the receiver performs inverse mapping, so that p_l can be obtained. k_a and k_b can be obtained by CCDM inverse mapping, thereby restoring the original data information.

4. Simulation Results

4.1. Rate Loss

This section numerically studies the rate loss of parallel distribution matcher based on CCDM. Take the probability distribution after quantization as $P_A = \{0.4, 0.3, 0.2, 0.1\}$ as an example to demonstrate the calculation process of the rate loss. When n = 10, C_{typ} = (4, 3, 2, 1), there are 49 complementary pairs (including C_{typ}) that satisfy the (5), the specific parameters of each complementary pair are shown in Table 1. Only the first 32 kinds of complementary pairs sorted by $k_a + k_b$ are listed in Table 1, because there are 49 kinds of complementary pairs, which are equivalent to 97 cases (C_{typ} is equivalent to only one case) and can index 6-bit binary data at most.

Table 1. The specific parameters of each complementary pair when n = 10, C_{typ} = (4,3,2,1).

Pair Number	C_l	$\overline{C_l}$	$k_a + k_b$	Pair Number	C_l	$\overline{C_l}$	$k_a + k_b$
1	(4, 3, 2, 1)	(4, 3, 2, 1)	13 + 13	17	(3, 2, 4, 1)	(5, 4, 0, 1)	13 + 10
2	(4, 3, 3, 0)	(4, 3, 1, 2)	12 + 13	18	(2, 4, 2, 1)	(6, 1, 2, 1)	12 + 11
3	(4, 4, 1, 1)	(4, 2, 3, 1)	12 + 13	19	(2, 3, 3, 2)	(6, 3, 1, 0)	14 + 9
4	(3, 4, 2, 1)	(5, 2, 2, 1)	13 + 12	20	(4, 4, 0, 2)	(4, 2, 4, 0)	11 + 11
5	(3, 3, 2, 2)	(5, 3, 2, 0)	14 + 11	21	(4, 5, 0, 1)	(4, 1, 4, 1)	10 + 12
6	(3, 3, 3, 1)	(5, 3, 1, 1)	13 + 12	22	(2, 5, 3, 0)	(6, 1, 1, 2)	11 + 11
7	(4, 4, 2, 0)	(4, 2, 2, 2)	11 + 13	23	(2, 3, 4, 1)	(6, 3, 0, 1)	19 + 9
8	(3, 5, 1, 1)	(5, 1, 3, 1)	12 + 12	24	(3, 5, 0, 2)	(5, 1, 4, 0)	11 + 10
9	(3, 4, 1, 2)	(5, 2, 3, 0)	13 + 11	25	(2, 5, 1, 2)	(6, 1, 3, 0)	12 + 9
10	(3, 4, 3, 0)	(5, 2, 1, 2)	12 + 12	25	(2, 4, 4, 0)	(6, 2, 0, 2)	11 + 10
11	(3, 2, 3, 2)	(5, 4, 1, 0)	14 + 10	27	(2, 2, 4, 2)	(6, 4, 0, 0)	14 + 7
12	(2, 4, 2, 2)	(6, 2, 2, 0)	14 + 10	28	(1, 5, 3, 1)	(7, 1, 1, 1)	12 + 9
13	(2, 4, 3, 1)	(6, 2, 1, 1)	13 + 11	29	(1, 4, 3, 2)	(7, 2, 1, 0)	13 + 8
14	(4, 5, 1, 0)	(4, 1, 3, 2)	10 + 13	30	(3, 6, 1, 0)	(5, 0, 3, 2)	9 + 11
15	(3, 5, 2, 0)	(5, 1, 2, 2)	11 + 12	31	(3, 1, 4, 2)	(5, 5, 0, 0)	13 + 7
16	(3, 3, 4, 0)	(5, 3, 0, 2)	12 + 11	32	(2 ,6, 1, 1)	(6, 0, 3, 1)	11 + 9

When the first 32 complementary pairs are chosen, p_l = 6, the average total number of indexable bits is 6 × 64 + 26 + 25 × 10 + 24 × 14 + 23 × 12 + 22 × 20 + 20 × 7 = 1852, the rate R = 1852/64/20 = 1.447 bit/symbol, and the rate of CCDM is 1.3 bit/symbol, increasing by 0.147 bit/symbol. When the first 17 complementary pairs are selected, p_l = 5 and the average total number of indexable bits is 5 × 32 + 26 + 25 × 10 + 24 × 14 + 23 × 7 = 933, resulting in a rate of R = 933/32/20 = 1.459 bit/symbol, an increase of 0.159 bit/symbol. When only the first three complementary pairs are selected, p_l = 2, the average total number of indexable bits is 2 × 4 + 26 + 25 × 3 = 109, and the rate R = 109/4 × 20 = 1.36 bit/symbol, increasing by 0.06 bit/symbol. In the worst case, the complementary pair may not be chosen, and the parallel CCDM is equivalent to a single CCDM.

Figure 3 shows rate loss over block length for the probability distribution as $P_A = \{0.4, 0.3, 0.2, 0.1\}$. The rates under different a are shown in Table 2. We observe that the parallel distribution matcher achieves a lower rate loss than CCDM for all block lengths n. When the block length n = 100, the rate of CCDM is 1.75 bit/symbol and the rate loss is 0.1, while when the parallel distribution matcher is n = 70, the maximum rate can reach 1.76 bit/symbol (p_l = 8) and the minimum is 1.74 bit/symbol (p_l = 4), saving 30% on block length n. When the block length of the parallel distribution matcher is n = 100, the maximum rate loss is 0.0847 bit/symbol and the minimum rate loss is 0.0693 bit/symbol, both better than CCDM's rate loss of 0.1 bit/symbol.

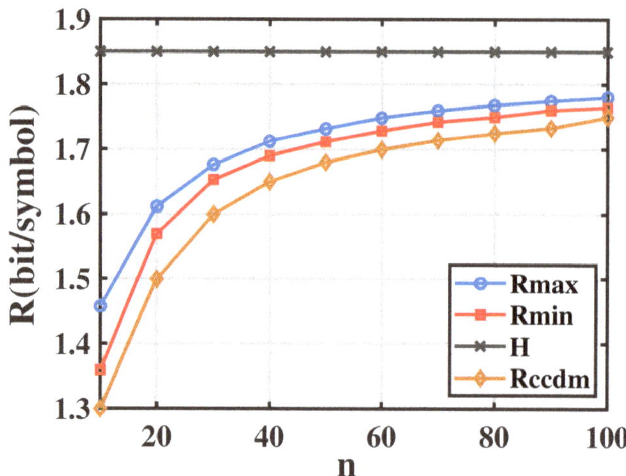

Figure 3. Rate over block length n for CCDM and parallel distribution matcher (Rmax and Rmin); the probability distribution is $P_A = \{0.4, 0.3, 0.2, 0.1\}$.

Table 2. Rate of CCDM and parallel DM under different block lengths n.

n	Parallel DM p_l = 8	Parallel DM p_l = 7	Parallel DM p_l = 6	Parallel DM p_l = 5	Parallel DM p_l = 4	CCDM
10	–	–	1.447	1.458	1.4375	1.3
20	–	–	1.612	1.606	1.589	1.5
30	–	1.677	1.673	1.665	1.653	1.6
40	1.713	1.711	1.706	1.699	1.69	1.65
50	1.732	1.73	1.725	1.72	1.712	1.68
60	1.7491	1.7462	1.7425	1.7352	1.7287	1.7
70	1.7604	1.7569	1.7521	1.747	1.743	1.7143
80	1.7685	1.7649	1.7611	1.7563	1.75	1.725
90	1.7749	1.7721	1.7679	1.7636	1.7604	1.733
100	1.7807	1.7767	1.7734	1.7702	1.7653	1.75

4.2. Generalized Mutual Information and BER Results

The Generalized mutual information (GMI) can be used to express the achievable information rate for bit-metric decoding. In this section, the generalized mutual information and bit error rate performance of the probability shaping signal using parallel distributed matcher are analyzed using the PAS scheme. The advantages of the proposed structure are demonstrated when compared with the PS-QAM signal using CCDM.

The GMI for bit-interleaved coded modulation is represented by (8) (Equation (13) of [3] and Equation (7) of [28]). In Monte Carlo simulations of N samples, GMI can be expressed as (9) (Equation (8) of [28]). Considering the rate loss of DM, the achievable information rate is shown in (10) (Equation (15) of [24]).

$$GMI(X;Y) = H(X) - \sum_{j=1}^{m} H(B_j|Y) \tag{8}$$

$$\begin{aligned} GMI &\approx \frac{1}{N}\sum_{k=1}^{N}[-\log_2 P_X(x_k)] - \frac{1}{N}\sum_{k=1}^{N}\sum_{i=1}^{m}[\log_2(1+e^{(-1)^{b_{k,i}\Lambda_{k,i}}}) \\ \Lambda_{k,i} &= \log\frac{\sum_{x\in\chi_1^i} p_{Y|X}(y_k|x)P_X(x)}{\sum_{x\in\chi_0^i} p_{Y|X}(y_k|x)P_X(x)} = \log\frac{p_{Y|B_i}(y_k|1)}{p_{Y|B_i}(y_k|0)} + \log\frac{P_{B_i}(1)}{P_{B_i}(0)} \end{aligned} \tag{9}$$

$$GMI_{DM} = GMI - R_{loss} \tag{10}$$

We consider a dual-polarization PS-64QAM modulation system with P(1, 3, 5, 7) = (0.4, 0.3, 0.2, 0.1), the baud rate per polarization is 28 GBaud. The simulation block diagram and parameters in VPI are shown in Figure 4. In Figure 5, GMI in bits per symbol is shown over OSNR in dB for PS-64QAM. Under the same OSNR, the greater n (n = 30, 50, 100) is, the smaller the rate loss is, and the larger the generalized mutual information is for the PS-64QAM signal utilizing the parallel distribution matcher. The OSNR required by the parallel distribution matcher is less than that using CCDM when n is 100 and the GMI is the same. When GMI is 4 bit/symbol, compared with the uniformly distributed 64QAM signal, the OSNR of the PS-64QAM signal using CCDM is reduced by 0.35 dB, and the OSNR of the PS-64QAM signal using the parallel distribution matcher is reduced by 0.47 dB, which is a 0.12 dB improvement compared with CCDM. When the block length of CCDM increases from n = 100 to n = 50,000, the OSNR only reduces by 0.33 dB; so, the parallel structure has a relatively large improvement of 0.12 dB.

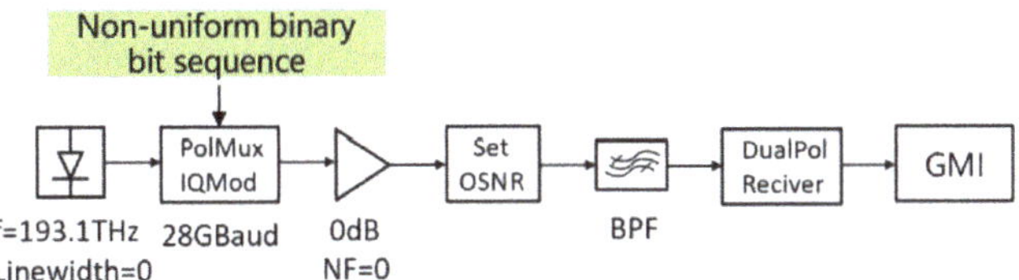

Figure 4. The GMI simulation block diagram and parameters in VPI.

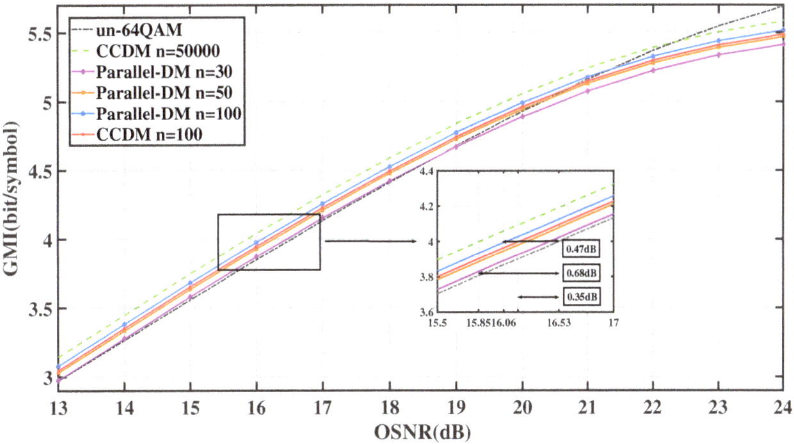

Figure 5. GMI in bit/symbol over OSNR in dB for bit-metric decoding and PS-64QAM. The inset zooms into the region around GMI = 4 bit/symbol, where the parallel distribution matcher of length n = 100 is 0.47 dB more power-efficient than uniform 64QAM, and is 0.12 dB more power-efficient than PS-64QAM using CCDM.

Figure 6 demonstrates the maximum and minimum GMI values for PS-64QAM signals using this parallel distribution matcher at different block lengths when the OSNR is 15, 16, and 17 dB. When OSNR = 15 dB, GMI = 3.655 bit/symbol; OSNR = 16 dB, GMI = 3.948 bit/symbol; and OSNR = 17 dB, GMI = 4.23 bit/symbol. The needed block length n of the parallel distribution matcher is only 60, although the required block length of the CCDM is 100; so, the block length can be saved by 40%. The matcher can attain the same performance as CCDM while using fewer blocks.

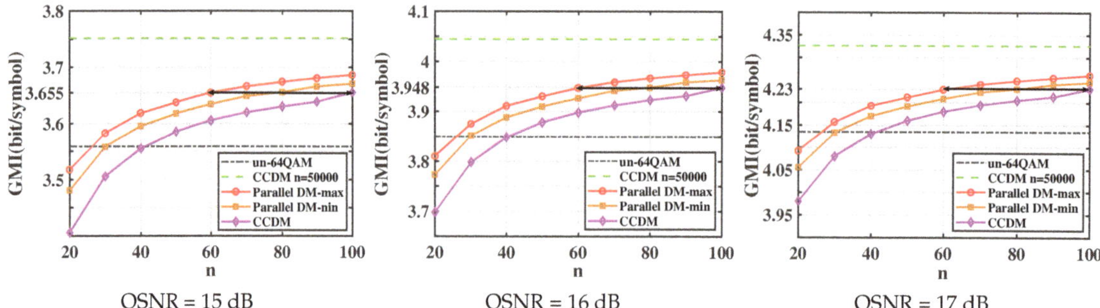

Figure 6. GMI in bit/symbol over block length n when OSNR = 15, 16, and 17 dB. When the GMI is the same, the parallel distribution matcher required block length n is only 60, while the required block length of CCDM is 100; so, the block length can be saved by 40%.

In terms of system bit error rate, the probability distribution of the final PS-QAM signal is the same whether utilizing the parallel distribution matcher or CCDM; the only difference is the number of PS-64QAM symbols generated. If n = 50 and the input bit sequence length is 58,128, the parallel structure can generate 33,600 PS-64QAM symbols, while the CCDM can generate 34,600 PS-64QAM symbols. When n = 100 and the input bit sequence length is 62,300, the parallel structure can generate 35,000 PS-64QAM symbols and the CCDM can generate 35,600 PS-64QAM symbols. Consequently, the bit error rate performance of the two is the same in the back-to-back situation.

When the transmission distance is long in the optical fiber transmission link, the amplifier spontaneous emission accumulates continuously and the number of symbols can have a slight impact on the bit error rate. Simulation block diagram and parameters of standard single-mode optical fiber transmission are shown in Figure 7. Figure 8 shows the BER performance under different transmission distances; the fiber is a standard single-mode fiber (SSMF) with α = 0.2 dB/km, $\gamma = 1.3$ $(\mathrm{W} \cdot \mathrm{km})^{-1}$, and D = 17 ps/nm/km; each span of length 100 km is followed by an Erbium-doped fiber amplifier with a noise figure of 3.8 dB. Laser phase noise and polarization mode dispersion are not included in the simulation as perfect compensation is assumed. The transmission distance of PS-64QAM signals using the paired optimized parallel distribution matcher is marginally greater than that of PS-64QAM signals using CCDM at the forward error correction threshold of −2.42 for n = 50 and n = 100.

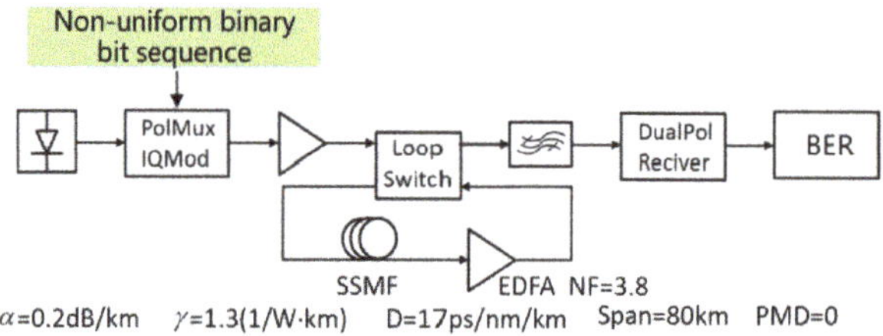

Figure 7. Simulation block diagram and parameters of standard single-mode optical fiber transmission.

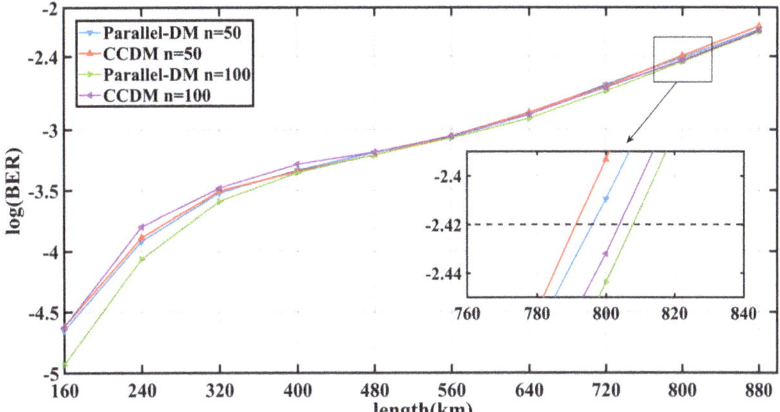

Figure 8. The BER performance over transmission distance in the standard single-mode fiber transmission link. The transmission distance of PS-64QAM signals using the parallel distribution matcher is slightly larger than that of PS-64QAM signals using CCDM.

5. Conclusions and Discussion

As a key part of probability shaping technology, the distribution matcher has significant importance and application value for research. At present, the widely used constant composition distribution matcher is a progressive optimization scheme. It is only when the length of the output symbols is infinity long that the rate loss of CCDM is zero. In addition, it adopts arithmetic coding, which is a highly serial coding method. It represents the input and output sequences, respectively, by dividing intervals between intervals. When the number of input and output sequences is large, the intervals to be divided become greater and the boundary between intervals becomes blurred, which is difficult to distinguish and easy to make mistakes when mapping. Therefore, it is necessary to propose a distribution matcher structure with good performance under short block length.

In this paper, we have proposed a novel parallel architecture distribution matcher, which is an improvement over conventional CCDM at short output block lengths. The output sequence has various compositions compared with the constant composition of the CCDM because of the parallel structure and pairwise optimization of the two CCDMs. As a result, the proposed structure has a lower rate loss when the output block lengths (n is less than 100) are the same and the output block lengths can be lowered by up to 30% with the same rate loss. In the simulation of an optical communication system of PS-64QAM signal, compared with a CCDM signal with the same generalized mutual information, the PS-64QAM signal employing this structure requires a lower OSNR, the block length can be reduced by 40%, and the transmission distance is increased. In detail, When GMI is 4 bit/symbol, compared with the uniformly distributed 64QAM signal, the OSNR of the PS-64QAM signal using CCDM is reduced by 0.35 dB, and the OSNR of the PS-64QAM signal using the parallel distribution matcher is reduced by 0.47 dB, which is a 0.12 dB improvement. The improvement in transmission distance is small, and it is expected to further improve the transmission distance in combination with digital signal processing technology, which will be our future research direction.

Other parallel distribution matcher schemes, such as parallel-amplitude distribution matcher [24], realize the function of symbol-level DM through multiple-bit-level DM. Therefore, M-1 bit-level DM is required to map sequences with M output amplitude values. For high-order QAM format, the number of bit-level DM required increases with the increase in modulation order, and the overall structure of the system is complex. In addition, the multiset-partition distribution [20], which also adopts the idea of pairwise optimization, needs to build a Huffman tree structure to index different complementary pairs and uses serial processing for the input data. The coding process is complex and error-prone. In contrast, the scheme proposed in this paper is a two-stage parallel structure. No matter how high the order of the QAM signal is, only two CCDMs are required. At the same time, the input data are processed in the two CCDMs after serial–parallel conversion. The receiver performs a corresponding parallel–serial conversion. There is no complex coding process; so, the structure is simple.

Author Contributions: Conceptualization, Y.Z. (Yao Zhang); methodology, Y.Z. (Yao Zhang) and H.W.; software and validation, Y.Z. (Yao Zhang); writing—original draft preparation, Y.Z. (Yao Zhang); supervision and project administration, H.W., Y.J. and Y.Z. (Yu Zhang); funding acquisition, H.W., Y.J. and Y.Z. (Yu Zhang). All authors have read and agreed to the published version of the manuscript.

Funding: This research was funded by National Key Research and Development Program of China (Grant No. 2019YFB1803601) and the National Natural Science Foundation of China (Grant No. 62021005).

Institutional Review Board Statement: Not applicable.

Informed Consent Statement: Not applicable.

Data Availability Statement: Not applicable.

Acknowledgments: The authors would like to thank the editors and anonymous reviewers for giving valuable suggestions that have helped to improve the quality of the manuscript.

Conflicts of Interest: The authors declare no conflict of interest.

References

1. Cisco, V.N.I. Global IP Traffic Forecast, 2017–2022. In *Cisco System*; Cisco: San Jose, CA, USA, 2018.
2. Cisco, U. *Cisco Annual Internet Report (2018–2023) White Paper*; Acessado Em; Cisco: San Jose, CA, USA, 2021.
3. Cho, J.; Winzer, P.J. Probabilistic constellation shaping for optical fiber communications. *J. Lightwave Technol.* **2019**, *37*, 1590–1607. [CrossRef]
4. Shannon, C.E. A mathematical theory of communication. *Bell Syst. Tech. J.* **1948**, *27*, 379–423. [CrossRef]
5. Kschischang, F.R.; Pasupathy, S. Optimal nonuniform signaling for Gaussian channels. *IEEE Trans. Inf. Theory.* **1993**, *39*, 913–929. [CrossRef]
6. Gallager, R.G. *Information Theory and Reliable Communication*; Wiley: New York, NY, USA, 1968; Volume 588.
7. Raphaeli, D.; Gurevitz, A. Constellation shaping for pragmatic turbo-coded modulation with high spectral efficiency. *IEEE Trans. Commun.* **2004**, *52*, 341–345. [CrossRef]
8. Yankov, M.; Forchhammer, S.; Larsen, K.J.; Christensen, L.P. Rate-adaptive constellation shaping for near-capacity achieving turbo coded BICM. In Proceedings of the 2014 IEEE International Conference on Communications (ICC), Sydney, Australia, 10–14 June 2014; pp. 2112–2117.
9. Kaimalettu, S.; Thangaraj, A.; Bloch, M.; McLaughlin, S.W. Constellation shaping using LDPC codes. In Proceedings of the 2007 IEEE International Symposium on Information Theory, Nice, France, 24–29 June 2007; pp. 2366–2370.
10. Le Goff, S.Y.; Sharif, B.S.; Jimaa, S.A. Bit-interleaved turbo-coded modulation using shaping coding. *IEEE Commun. Lett.* **2005**, *9*, 246–248. [CrossRef]
11. Böcherer, G.; Mathar, R.; Jimaa, S.A. Operating LDPC codes with zero shaping gap. In Proceedings of the 2011 IEEE Information Theory Workshop (ITW), Paraty, Brazil, 16–20 October 2011; pp. 330–334.
12. Böcherer, G.; Steiner, F.; Schulte, P. Bandwidth efficient and rate-matched low-density parity-check coded modulation. *IEEE Trans. Commun.* **2015**, *63*, 4651–4665. [CrossRef]
13. Böcherer, G.; Schulte, P.; Steiner, F. Probabilistic shaping and forward error correction for fiber-optic communication systems. *J. Lightwave Technol.* **2019**, *37*, 230–244. [CrossRef]
14. Ghazisaeidi, A.; de Jauregui Ruiz, I.F.; Rios-Müller, R.; Schmalen, L.; Tran, P.; Brindel, P.; Renaudier, J. Advanced C+ L-band transoceanic transmission systems based on probabilistically shaped PDM-64QAM. *J. Lightwave Technol.* **2017**, *35*, 1291–1299. [CrossRef]
15. Idler, W.; Buchali, F.; Schmalen, L.; Lach, E.; Braun, R.P.; Böcherer, G.; Steiner, F. Field trial of a 1 Tb/s super-channel network using probabilistically shaped constellations. *J. Lightwave Technol.* **2017**, *35*, 1399–1406. [CrossRef]
16. Cho, J.; Chen, X.; Chandrasekhar, S.; Raybon, G.; Dar, R.; Schmalen, L.; Grubb, S. Trans-atlantic field trial using high spectral efficiency probabilistically shaped 64-QAM and single-carrier real-time 250-Gb/s 16-QAM. *J. Lightwave Technol.* **2017**, *36*, 103–113. [CrossRef]
17. Kong, M.; Yu, J.; Chien, H.C.; Wang, K.; Li, X.; Shi, J.; Chen, Y. WDM Transmission of 600G Carriers over 5600 km with Probabilistically Shaped 16QAM at 106 Gbaud. In Proceedings of the Optical Fiber Communication Conference (OFC) 2019, San Diego, CA, USA, 3–7 March 2019; pp. 1–3.

18. Kong, M.; Wang, K.; Ding, J.; Zhang, J.; Li, W.; Shi, J.; Yu, J. 640-Gbps/carrier WDM transmission over 6,400 km based on PS-16QAM at 106 Gbaud employing advanced DSP. *J. Lightwave Technol.* **2020**, *39*, 55–63. [CrossRef]
19. Schulte, P.; Böcherer, G. Constant composition distribution matching. *IEEE Trans. Inf. Theory.* **2015**, *62*, 430–434. [CrossRef]
20. Fehenberger, T.; Millar, D.S.; Koike-Akino, T.; Kojima, K.; Parsons, K. Multiset-partition distribution matching. *IEEE Trans. Commun.* **2018**, *67*, 1885–1893. [CrossRef]
21. Fehenberger, T.; Millar, D.S.; Koike-Akino, T.; Kojima, K.; Parsons, K.; Griesser, H. Huffman-coded sphere shaping and distribution matching algorithms via lookup tables. *J. Lightwave Technol.* **2020**, *38*, 2826–2834. [CrossRef]
22. Pikus, M.; Xu, W. Bit-level probabilistically shaped coded modulation. *IEEE Commun. Lett.* **2017**, *21*, 1929–1932. [CrossRef]
23. Koganei, Y.; Sugitani, K.; Nakashima, H.; Hoshida, T. Optimum bit-level distribution matching with at most O (N3) implementation complexity. In Proceedings of the Optical Fiber Communication Conference (OFC) 2019, San Diego, CA, USA, 3–7 March 2019; p. M4B-4.
24. Fehenberger, T.; Millar, D.S.; Koike-Akino, T.; Kojima, K.; Parsons, K. Parallel-amplitude architecture and subset ranking for fast distribution matching. *IEEE Trans. Commun.* **2020**, *68*, 1981–1990. [CrossRef]
25. Yoshida, T.; Karlsson, M.; Agrell, E. Hierarchical distribution matching for probabilistically shaped coded modulation. *J. Lightwave Technol.* **2019**, *37*, 1579–1589. [CrossRef]
26. Goossens, S.; Van der Heide, S.; Hout, M.V.D.; Amari, A.; Gultekin, Y.C.; Vassilieva, O.; Kim, I.; Ikeuchi, T.; Willems, F.M.J.; Alvarado, A.; et al. First experimental demonstration of probabilistic enumerative sphere shaping in optical fiber communications. In Proceedings of the 24th OptoElectronics and Communications Conference (OECC) and 2019 International Conference on Photonics in Switching and Computing (PSC), Fukuoka, Japan, 7–11 July 2019; pp. 1–3.
27. Gültekin, Y.C.; Fehenberger, T.; Alvarado, A.; Willems, F.M.J.Probabilistic shaping for finite blocklengths: Distribution matching and sphere shaping. *J. Entropy.* **2020**, *22*, 581. [CrossRef] [PubMed]
28. Fehenberger, T.; Alvarado, A.; Böcherer, G.; Hanik, N. On probabilistic shaping of quadrature amplitude modulation for the nonlinear fiber channel. *J. Lightwave Technol.* **2016**, *34*, 5063–5073. [CrossRef]

Article

Numerical Study on a Bound State in the Continuum Assisted Plasmonic Refractive Index Sensor

Shulin Tang [1,2,3], Chang Chang [1], Peiji Zhou [1] and Yi Zou [1,4,*]

[1] School of Information Science and Technology, Shanghai Tech University, Shanghai 201210, China; tangshl@shanghaitech.edu.cn (S.T.); changch@shanghaitech.edu.cn (C.C.); zhoupj@shanghaitech.edu.cn (P.Z.)
[2] Shanghai Institute of Microsystem and Information Technology, Chinese Academy of Sciences, Shanghai 200050, China
[3] University of Chinese Academy of Sciences, Beijing 100049, China
[4] Shanghai Engineering Research Center of Energy Efficient and Custom AI IC, Shanghai 201210, China
* Correspondence: zouyi@shanghaitech.edu.cn

Abstract: Plasmonic sensors have attracted intensive attention due to their high sensitivity. However, due to intrinsic metallic loss, plasmonic sensors usually have a large full width at half maximum (FWHM) that limits the wavelength resolution. In this paper, we numerically investigate and propose a dielectric grating-assisted plasmonic device, leveraging the bound states in the continuum (BIC) effect to suppress the FWHM of the resonance. We initiate quasi-SP-BIC modes at 1559 nm and 1905 nm wavelengths by slightly tilting the incident angle at 2° to break the symmetry, featuring a narrow linewidth of 1.8 nm and 0.18 nm at these two wavelengths, respectively. Refractive index sensing has also been investigated, showing high sensitivity of 938 nm/RIU and figure of merit (FOM) of 521/RIU at 1559 nm and even higher sensitivity of 1264 nm/RIU and FOM of 7022/RIU at 1905 nm.

Keywords: bound states in the continuum; plasmonic; sensor

Citation: Tang, S.; Chang, C.; Zhou, P.; Zou, Y. Numerical Study on a Bound State in the Continuum Assisted Plasmonic Refractive Index Sensor. *Photonics* **2022**, *9*, 224. https://doi.org/10.3390/photonics9040224

Received: 28 January 2022
Accepted: 2 March 2022
Published: 28 March 2022

Publisher's Note: MDPI stays neutral with regard to jurisdictional claims in published maps and institutional affiliations.

1. Introduction

Over the past two decades, there has been extensive research on plasmonic devices due to the capability to break the optical diffraction limit and confine light in sizes that are much smaller than the diffraction limit, thus enhancing the electric field [1,2]. Because of the surface wave nature, this type of device is usually very sensitive to surrounding refractive index changes [3–8], making them potential candidates for sensing application. A recent review paper summarized the state of the art in this area [9]. However, the intrinsic loss from the materials limits the linewidth of the device resonance and hinders further development. For example, interacting between impinging light and lossy materials will generate a thermal issue that modifies the surrounding medium and changes the sensitivity. One possible way to address this problem is to use BICs to mitigate the issue [10]. A figure-of-merit (FOM), defined as S/FWHM (S denotes the sensitivity, and FWHM denotes the full width at half maximum), is often adopted to characterize a sensor performance. An ideal sensor would possess a larger FOM, namely high sensitivity and a small FWHM. Therefore, an approach to reduce the FWHM of plasmonic sensors while maintaining their sensitivity is highly desired.

BIC, which is referred to as embedded eigenvalues or embedded trapped modes, was first proposed by von Neumann and Wigner in 1929 [11]. It is a localized state of an open structure with access to radiation channels, yet it remains highly confined with, in theory, an infinite lifetime and Q-factor. Therefore, it is considered an important approach for designing high-Q optical resonators [12–15] and has been applied to lasers [16–19] and sensors [20–22]. Recently, Azzam et al. [23] proposed a hybrid plasmonic-photonic structure that possesses two types of BICs, namely symmetry-protected BICs (SP-BICs) and Friedrich–Wintgen BICs (FW-BICs), of which the former is caused by the symmetry incompatibility between the local modes and the continuum spectrum and the latter is induced by the destructive interference of two scattering channels.

In this paper, we numerically study a BIC-assisted plasmonic sensor that leverages the BIC effect to suppress the linewidth of the resonance. A quasi-SP-BIC mode at 1559 nm can be initiated by slightly tilting the incident angle to break the symmetry, featuring a narrow FWHM of 1.8 nm. The device possesses high sensitivity (~938 nm/RIU) (RIU denotes refractive index unit) and high FOM (~521/RIU) characteristics in refractive index sensing. By tuning the operating wavelength to 1905 nm, our device exhibits a higher sensitivity of ~1264 nm/RIU, a narrower FWHM of 0.18 nm, and a better FOM of ~7022/RIU. The devices also feature a 0.1 nm/(g/L) sensitivity for glucose solution detection at 1559 nm and a 0.14 nm/(g/L) sensitivity at 1905 nm.

2. Design of a BIC-Assisted Plasmonic Sensor

The schematic of the proposed device is shown in Figure 1a. It consists of a one-dimensional aluminum nitride (AlN) grating on a metal–glass substrate. The momentum matching condition between the excited resonance mode and the incident light is [24]:

$$k_{sp} = \frac{2\pi}{\lambda} n_s sin\theta + m\frac{2\pi}{a} = \frac{2\pi}{\lambda}\sqrt{\frac{\varepsilon_m {n_s}^2}{\varepsilon_m + {n_s}^2}} \tag{1}$$

where k_{sp} is the momentum of the surface plasmon, λ is the free-space wavelength, n_s is the refractive index of the surrounding medium, ε_m is the dielectric constant of the metal layer, θ is the angle of incidence, a is the grating pitch, and m is the grating diffraction order.

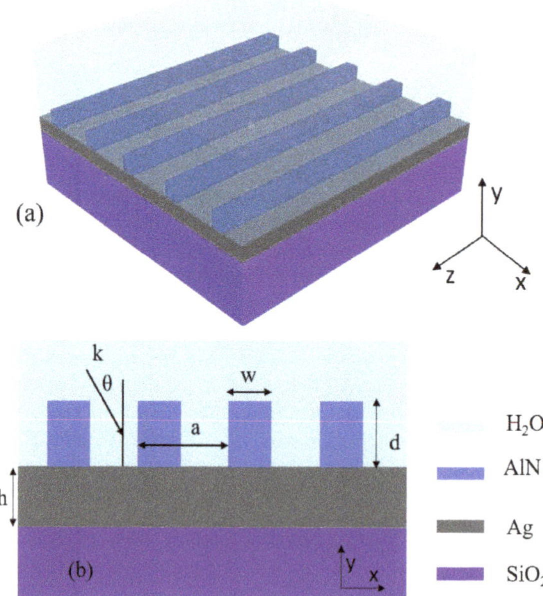

Figure 1. (**a**) Schematic of the proposed device, consisting of one-dimensional aluminum nitride (AlN) grating on Ag-SiO2 substrate. (**b**) Cross-section of the device. Here, the width and height of the grating are w = 320 nm and d = 440 nm, respectively, the period is a = 1070 nm, and the thickness of the silver film is h = 100 nm.

In our device, we choose silicon dioxide as the substrate and silver as the metal layer for its low-loss property in the near-infrared range among all the metals [25]. We pick AlN as the grating material for its transparency in the near-infrared, which can effectively reduce the device loss and lead to a narrower resonance peak. For our simulation, we fix the refractive index of silicon dioxide at 1.45 for its negligible variations. The refractive indices of silver and AlN are from Refs [26,27], respectively. We set water as the surrounding medium for sensor design and fix its refractive index at 1.33. Figure 1b shows the cross-section of the device. The AlN grating with the width, w = 320 nm, is periodically arranged in the x-direction with a period, a = 1070 nm. The thicknesses of the grating, d, and silver film, h, are 440 nm and 100 nm, respectively. The value of the geometric parameter shown in Figure 1b is listed in Table 1.

Table 1. Value of the Geometric Parameter Shown in Figure 1b.

w (nm)	d (nm)	a (nm)	h (nm)	Incident Angle (deg)
320	440	1070	100	2

We employ the finite difference time domain (FDTD) method to calculate the reflection of the proposed device. Due to the invariance in the z-direction of our device, for convenience, 2D-FDTD is chosen. We simulate the device using a non-uniform grid with a minimum grid size of 1 nm. The perfectly matched boundary condition and periodic Bloch boundary condition are selected for the y-direction and the x-direction, respectively. We use TM polarized light as the incident light to excite the plasmonic mode at the metal-dielectric interface.

The calculated reflected spectra for the wavelength range of 1350–1650 nm under different incident angles are plotted in Figure 2a. Two counter-propagating plasmonic modes, SP [+1] and SP [−1], form standing waves, resulting in the formation of symmetric and anti-symmetric modes with different energy bands as shown in the insets of Figure 2b. The bandgap between the two energy bands covers the wavelength region of 1475–1540 nm, and the width indicates the strength of the interaction between the two modes. It is worth noting that, in the upper band, there is a discontinuous region, circled by the red dash line, at the wavelength ~1559 nm when the light is normally incident, which does not exist in the lower band. To better illustrate this phenomenon, we plot the reflected spectral when the incident angles are 0° and 2° in Figure 2b. Compared with the normal incidence, a clear Fano resonance, marked by a red dashed square, is observed around 1559 nm when the device is illuminated with a 2° incident angle, indicating an SP-BIC exists in the device. The corresponding magnetic field distribution is shown in the inset of Figure 2b, where the magnetic field is antisymmetric to the central line. Since the incident plane wave is symmetric when the wavevector is normal to the plane. Additionally, the reflected wave and transmitted wave are supposed to be plane waves with no diffraction order other than the 0th order due to the small lattice constant. The different types of symmetry prevent coupling between the SP-BIC mode and the output wave, granting the strong localization of this mode. Since the incident wave cannot couple to the other modes at this wavelength, most of the energy is reflected, resulting in the high reflective coefficient, which explains the discontinuity of the upper band in Figure 2a. To excite a quasi-SP-BIC mode, we can slightly tilt the incident angle to break the symmetry and provide a certain amount of mode overlap with the quasi-SP-BIC mode.

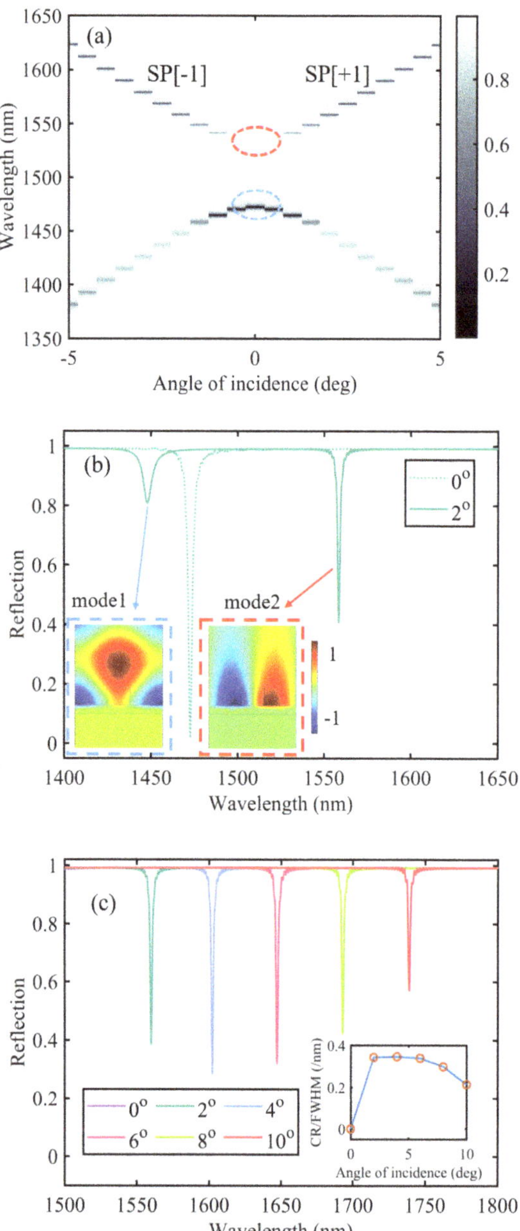

Figure 2. (**a**) The reflected spectra for different incident angles. A discontinuity is observed at the center of the red circle, indicating the existence of an SP-BIC. (**b**) The reflection spectrum when the incident angles are 0° and 2°, respectively. The insets show the magnetic field distribution for the corresponding resonances circled by the red and blue lines in Figure 2a. (**c**) The reflection spectral for different incident angles. The inset shows the value of the CR/FWHM for different incident angles.

Compared with the anti-symmetric mode in the upper band, the lower band features symmetric mode distribution, as shown in the blue dashed square in Figure 2b. Due to the same symmetry between the incident light and the mode, even normal incidence can excite the resonance, which results in the continuity of the lower band in Figure 2a.

In Figure 2b, we also observe the different linewidths of the two modes. As seen from the insets, compared with the symmetric mode, the anti-symmetric mode is better confined in the device, which results in a reduced radiation loss. Therefore, the anti-symmetric mode (FWHM = 1.8 nm) has a narrower resonance peak than the symmetric one (FWHM = 5.7 nm). Although exciting the SP-BIC is hard, we can leverage the quasi-SP-BIC mode to suppress the resonance linewidth and significantly improve the device sensing performance.

We also study the property of the quasi-SP-BIC mode by gradually tilting the incident angle to deviate from zero-degree, as shown in Figure 2c. As the incident angle gradually increases, the resonance contrast ratio (CR), which refers to the reflection difference between the off-resonance and the on-resonance, becomes larger, indicating more energy couples into the quasi-SP-BIC mode due to increasing mode overlap. However, this process is associated with increased radiation loss due to the deviation from the SP-BIC condition, which, in turn, widens the FWHM. Therefore, we make a trade-off between the FWHM and the resonance CR by calculating the value of CR/FWHM, as shown in the inset of Figure 2c, and choose 2° as the final incident angle for its biggest value of CR/FWHM.

3. Results

3.1. Sensitivity Characterization and Refractive Index Sensing

To evaluate the device sensing performance, we simulate the response of the device under different environmental mediums with a refractive index range from 1.33 to 1.57. During the calculation, we fixed the grating period, height, and width to 1070 nm, 440 nm, and 320 nm, respectively, and the silver thickness to 100 nm. Here, we set the incident angle to 2° to slightly break its symmetry and enter the quasi-SP-BIC mode. Figure 3a shows multiple reflected spectra when the surrounding refractive index gradually increases. The quasi-SP-BIC mode red-shifts from 1559 nm to 1784 nm while the other resonance peak red-shifts from 1447 nm to 1647 nm. Linear fittings in Figure 3b indicate a linear resonance shift for both the two modes, of which the former has a sensitivity of ~938 nm/RIU and the latter's sensitivity is 833 nm/RIU. Considering the sharpness of the resonance, we obtain the FOM of the quasi-SP-BIC mode about 521/RIU, which is also the FOM of our device. In the following part, we will focus our discussion on the quasi-SP-BIC mode for its higher FOM.

Figure 3c presents the simulated resonance shift corresponding to different glucose concentrations in water with the incident angle at 2°. The refractive index of the glucose solution is taken from Ref. [28]. The resonance red-shifts linearly as the glucose concentration increases, showing 0.1 nm/(g/L).

3.2. Performance Change with the Thickness of the Silver Layer

In our device, we choose silver to generate the plasmonic mode. Its thickness plays an important role in the device's performance. It has been proved that when the metal layer is thinner, parts of the energy are radiated into the substrate, resulting in a larger radiation loss, which gradually stabilizes when the thickness of the metal layer approaches the skin depth of the metal [29]. We calculate the reflected spectra of the quasi-SP-BIC under different silver layer thicknesses in Figure 4. During the calculation, all other parameters are fixed (a = 1070 nm, w = 320 nm, d = 440 nm, n_s = 1.33, θ = 2°). The thickness of the silver layer could slightly affect the position and the linewidth of the resonance. When the thickness changes from 60 nm to 120 nm with a step of 20 nm, the resonance wavelengths are 1558.64 nm, 1558.42 nm, 1558.56 nm, and 1558.47 nm, and the corresponding FWHM, shown in the inset of Figure 4, are 2 nm, 1.8 nm, 1.8 nm, and 1.8 nm. When the silver layer is thicker than 80 nm, the FWHM becomes stable, indicating no further substrate leakage occurs. In addition, the resonance CR stabilizes when the thickness is greater than 80 nm. Considering the FWHM and the resonance CR, a silver film with a thickness larger than

80 nm is preferred. We thus pick 100 nm as our silver thickness to provide a sufficient buffer for device implementation.

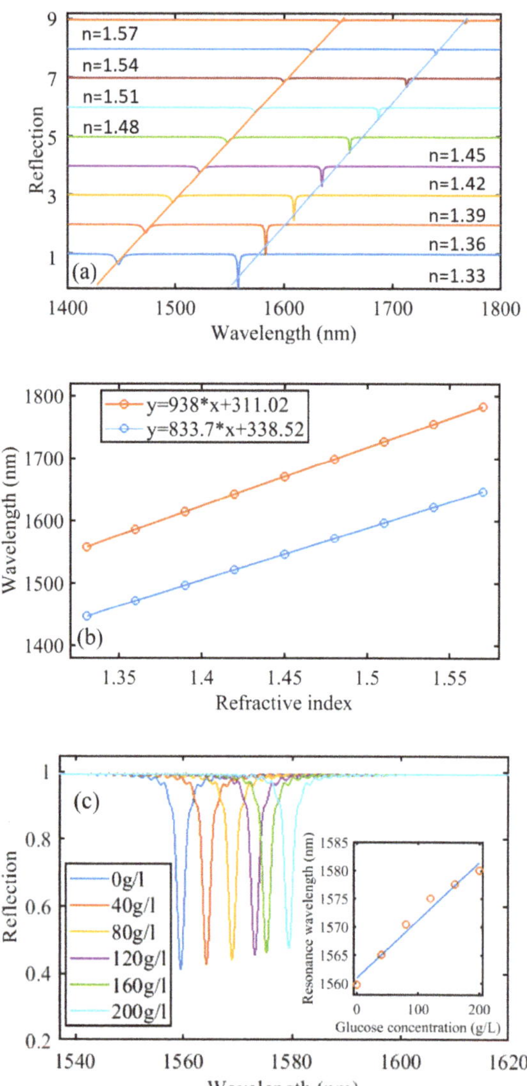

Figure 3. (**a**) The reflected spectra for different surrounding mediums. (**b**) The resonance wavelengths shift as a function of the surrounding index variation. The linear fittings show 833 nm/RIU and 938 nm/RIU for Mode1 and Mode2, respectively. (**c**) The reflected spectra for different concentrations of the glucose solution. The inset shows the resonance wavelength shifts with the glucose solution concentration variation.

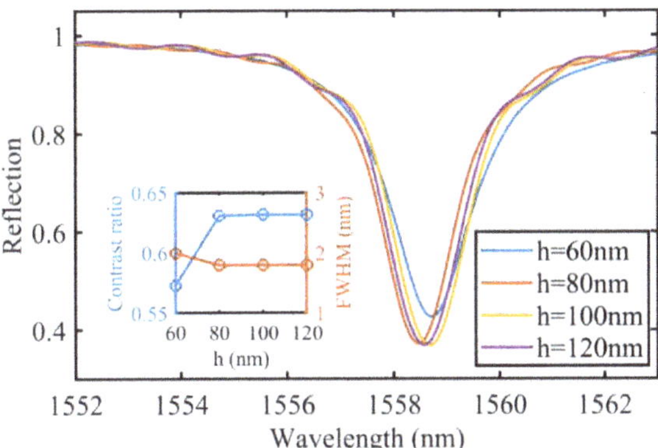

Figure 4. The reflected spectra for different thicknesses of the silver layer. The inset shows the resonance CR and the FWHM change with the silver layer thickness.

3.3. Performance Change with the Grating Variables

The AlN grating is also crucial for our device. There are multiple parameters in a grating, e.g., the grating height, width, and period, that would affect the device's performance. In this section, we provide a detailed analysis of these parameters on the performance of the sensor. Note that in this section, the silver thickness is fixed at 100 nm.

First, we calculate the reflected spectra for different grating heights with fixed grating width (w = 320 nm), period (a = 1070 nm), and incident angle (θ = 2°). Figure 5a shows that when the grating height gradually thickens from 400 nm to 480 nm, the resonance moves to the longer wavelength without significant distortion of its shape. The corresponding FWHM of the resonant peak is around 1.8 nm during the parameter scanning. We also do refractive index sensitivity characterization, showing the sensitivity is within the range of 920 nm/RIU to 960 nm/RIU. Therefore, a thinner grating possesses a higher sensitivity and thus a higher FOM, as shown in the inset of Figure 5a. However, from Figure 5a, we observe a slight decrease in the resonance CR for the grating height is smaller than 440 nm. Considering the overall performance, we pick 440 nm as our grating height to provide a higher FOM without sacrificing the CR of the resonance. This also provides a reference for our future design.

Next, we plot the reflected spectra for different grating widths with fixed grating height (d = 440 nm), period (a = 1070 nm), and incident angle (θ = 2°) in Figure 5b. We find a red-shifted and slightly broadened resonant peak as we widen the grating width, i.e., increasing the duty cycle. The FWHM increases from 1.6 nm for the 280 nm wide grating to 2 nm for the 360 nm wide one, while the sensing sensitivity monotonically drops from 958 nm/RIU for the 280 nm wide grating to 923 nm/RIU for the 360 nm wide grating. The corresponding FOM also monotonically drops from 598/RIU to 461/RIU, as shown in the inset of Figure 5b, indicating a better sensing performance for the grating with a lower duty cycle. However, the inset of Figure 5b also shows an increasing resonance CR for wider grating teeth. Therefore, by taking a trade-off, we pick 320 nm as the grating width.

We also simulated the reflected spectra for different grating periods (from 970 nm to 1170 nm with a 50 nm step) and kept all other parameters fixed (w = 320 nm, d = 440 nm, and θ = 2°). As shown in Figure 5c, when the period increases, the corresponding FWHM slightly drops from 1.9 nm to 1.7 nm, and the sensitivity increases from 826 nm/RIU to 1060 nm/RIU, providing a monotonically increasing FOM from 434/RIU to 623/RIU. In the meantime, we also notice a decreasing resonance CR as we increase the period. Therefore, we trade-off these factors and select the period a = 1070 nm for our design.

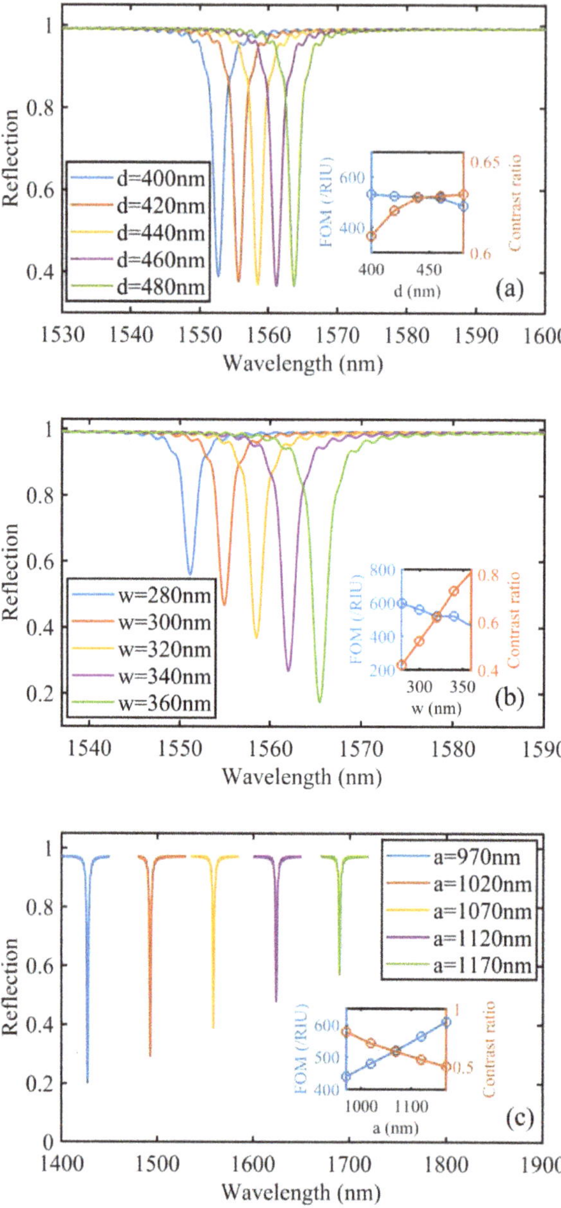

Figure 5. The reflected spectra for the grating with (**a**) different heights, (**b**) different widths, and (**c**) different grating periods. Insets show the corresponding FOM and the resonance CR versus the grating (**a**) height, (**b**) width, and (**c**) period.

3.4. Performance Change with the Grating Shape Deviation

We also study the device tolerance by considering the fabrication-induced sidewall angle deviations. In reality, it is common to have a trapezoid cross-section similar to the inset of Figure 6. In this section, we fix all the parameters same as in Section 2, namely a = 1070 nm, d = 440 nm, w = 320 nm, h = 100 nm, $\theta = 2°$, and n_s = 1.33, and scan the Δw

from −50 nm to 50 nm with a step of 25 nm. We plot the reflected spectra for different values of Δw in Figure 6. As Δw increases from −50 nm to 50 nm, similar to widening the grating width, the resonance shifts to the longer wavelength, and the CR gradually increases. The corresponding sensitivity decreases from 945 nm/RIU to 924 nm/RIU. During the whole process, the FWHM does not vary too much. It is fair to say that the device shows a sufficient tolerance for certain fabrication-induced geometric errors.

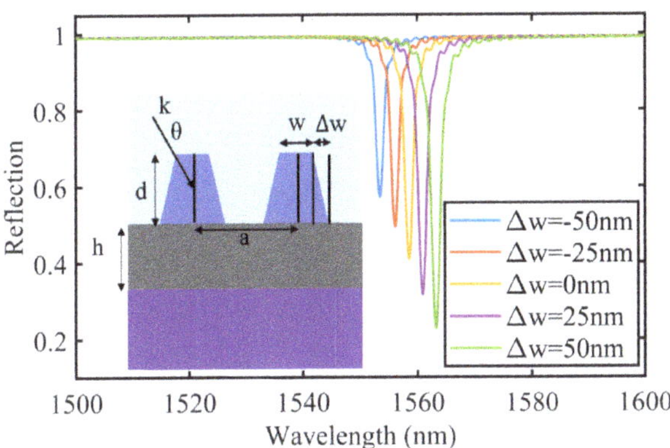

Figure 6. The reflected spectra for different Δw. The inset shows the cross-section of the device.

4. Discussion

According to Equation (1), the resonance wavelength is proportional to the grating period. Further increasing the period may move the operating wavelength towards a longer wavelength. On the other hand, if we define the complex refractive index of silver as $n_m = n + ik$, for a plasmonic device, the quality factor is [30]:

$$Q_{sp} = \frac{k^3}{2n} \tag{2}$$

A larger quality factor indicates a lower loss of a device. Take 1559 nm and 1905 nm as an example. Even though moving from 1559 nm to about 1905 nm increases the imaginary part of the complex refractive index, k, from 11.439 to 13.862, it also raises the real part, n, from 0.1453 to 0.23119. Therefore, the Q_{sp} indeed increases from about 5150 to about 5760, implying a lower device loss when working at a longer wavelength. In the meantime, a resonance working at a longer wavelength shifts more than the one at a shorter wavelength.

We thus increase the period to shift the resonance peak of the quasi-SP-BIC from 1559 nm to a longer wavelength. Then, we optimize the device parameters as a = 1350 nm, w = 400 nm, d = 240 nm, and h = 100 nm. With incident angle θ = 2°, we apply glucose solution with different concentrations to the device, showing 1264 nm/RIU, corresponding to 0.14 nm/(g/L) for glucose solution, and 0.18 nm FWHM in Figure 7. The sensitivity is more than 30% higher than its counterpart at 1559 nm, while the FWHM is one order narrower than the 1559 nm one, leading to a much higher FOM of ~7022/RIU. We compare the performance of our devices with other recently reported plasmonic sensors in Table 2.

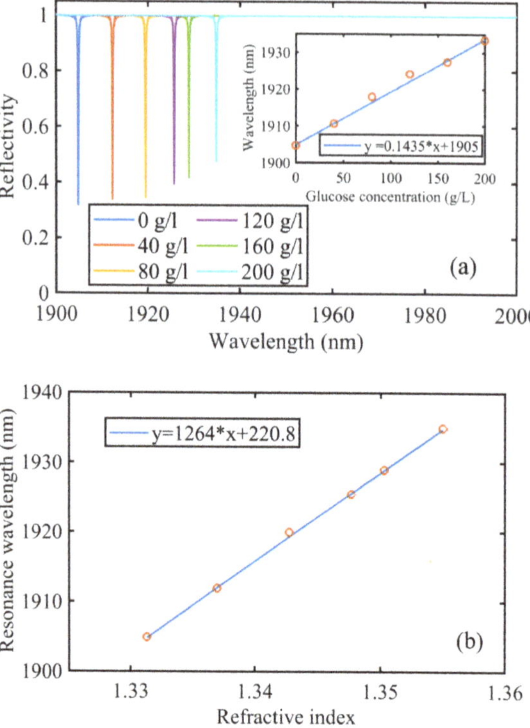

Figure 7. (**a**) The reflected spectra for different concentrations of the glucose solution. The inset shows resonance wavelength shifts with the glucose solution concentration variation. (**b**) The resonance wavelengths shift with the glucose solution concentration change-induced surrounding index variation. The linear fitting shows 1264 nm/RIU.

Table 2. Performance comparison.

Ref.	Sensitivity (nm/RIU)	FWHM (nm)	FOM (/RIU)
Sharma et al. [31]	461.53	14.8	31.18
Lu et al. [32]	497.83	0.904	551
Chau et al. [33]	1200	45	26.67
Sun et al. [34]	526.0	7.2	73.10
Li et al. [35]	404.295	8.04	50.30
Sreekanth et al. [36]	30,000	50.8	590
He et al. [37]	815	360.6	2.26
This paper @1559 nm	938	1.8	521
This paper @1905 nm	1264	0.18	7022

5. Conclusions

In summary, we propose a grating-assisted plasmonic refractive index sensor. Combining the quasi-SP-BIC effect and the surface wave nature, our device features a small FWHM and high sensitivity, resulting in FOM of 521/RIU at 1559 nm and 7022/RIU at 1905 nm. We also do glucose solution sensing, showing a 0.1 nm/(g/L) sensitivity at 1559 nm and a 0.14 nm/(g/L) sensitivity at 1905 nm. The performance of the proposed scheme is significantly better than the other grating-based plasmonic sensors reported earlier. Our approach provides a feasible way for designing a high-performance sensor with narrow FWHM and high sensitivity.

Author Contributions: Conceptualization, S.T. and P.Z.; methodology, S.T.; software, S.T. and C.C.; validation, S.T.; formal analysis, S.T.; investigation, S.T.; resources, S.T. and Y.Z.; data curation, S.T.; writing—original draft preparation, S.T.; writing—review and editing, Y.Z.; visualization, S.T.; supervision, Y.Z.; project administration, Y.Z.; funding acquisition, Y.Z. All authors have read and agreed to the published version of the manuscript.

Funding: The research was sponsored by the National Natural Science Foundation of China (NSFC) (61705099) and the Natural Science Foundation of Shanghai (21ZR1443100).

Institutional Review Board Statement: Not applicable.

Informed Consent Statement: Not applicable.

Data Availability Statement: Not applicable.

Conflicts of Interest: The funders had no role in the design of the study; in the collection, analyses, or interpretation of data; in the writing of the manuscript; or in the decision to publish the results.

References

1. Reather, H. Surface Plasmons on Smooth and Rough Surfaces and on Gratings. *Springer Tracts Mod. Phys.* **1988**, *111*, 345–398.
2. Pandey, A.K.; Sharma, A.K.; Basu, R. Fluoride glass-based surface plasmon resonance sensor in infrared region: Performance evaluation. *J. Phys. D Appl. Phys.* **2017**, *50*, 185103. [CrossRef]
3. Sharma, S.; Kumari, R.; Varshney, S.K.; Lahiri, B. Optical biosensing with electromagnetic nanostructures. *Rev. Phys.* **2020**, *5*, 100044. [CrossRef]
4. Suido, Y.; Yamamoto, Y.; Thomas, G.; Ajiki, Y.; Kan, T. Extension of the measurable wavelength range for a near-infrared spectrometer using a plasmonic Au grating on a si substrate. *Micromachines* **2019**, *10*, 403. [CrossRef]
5. Chen, W.; Kan, T.; Ajiki, Y.; Matsumoto, K.; Shimoyama, I. NIR spectrometer using a Schottky photodetector enhanced by grating-based SPR. *Opt. Express* **2016**, *24*, 25797–25804. [CrossRef]
6. Dhawan, A.; Gerhold, M.D.; Muth, J.F. Plasmonic structures based on subwavelength apertures for chemical and biological sensing applications. *IEEE Sens. J.* **2008**, *8*, 942–950. [CrossRef]
7. Fore, S.; Yuen, Y.; Hesselink, L.; Huser, T. Pulsed-interleaved excitation FRET measurements on single duplex DNA molecules inside C-shaped nanoapertures. *Nano Lett.* **2007**, *7*, 1749–1756. [CrossRef]
8. Zaman, M.A.; Padhy, P.; Hesselink, L. Solenoidal optical forces from a plasmonic Archimedean spiral. *Phys. Rev. A* **2019**, *100*, 013857. [CrossRef]
9. Palermo, G.; Sreekanth, K.V.; Maccaferri, N.; Lio, G.E.; Nicoletta, G.; De Angelis, F.; Hinczewski, M.; Strangi, G. Hyperbolic dispersion metasurfaces for molecular biosensing. *Nanophotonics* **2021**, *10*, 295–314. [CrossRef]
10. Ferraro, A.; Lio, G.E.; Hmina, A.; Palermo, G.; Djouda, J.M.; Maurer, T.; Caputo, R. Tailoring of plasmonic functionalized metastructures to enhance local heating release. *Nanophotonics* **2021**, *10*, 3907–3916. [CrossRef]
11. Von Neumann, J.; Wigner, E. On some peculiar discrete eigenvalues. *Phys. Z.* **1929**, *30*, 465–467.
12. Plotnik, Y.; Peleg, O.; Dreisow, F.; Heinrich, M.; Nolte, S.; Szameit, A.; Segev, M. Experimental observation of optical bound states in the continuum. *Phys. Rev. Lett.* **2011**, *107*, 183901. [CrossRef] [PubMed]
13. Fan, K.; Shadrivov, I.V.; Padilla, W.J. Dynamic bound states in the continuum. *Optica* **2019**, *6*, 169–173. [CrossRef]
14. Hsu, C.W.; Zhen, B.; Stone, A.D.; Joannopoulos, J.D.; Soljačić, M. Bound states in the continuum. *Nat. Rev. Mater.* **2016**, *1*, 1–13. [CrossRef]
15. Molina, M.I.; Miroshnichenko, A.E.; Kivshar, Y.S. Surface bound states in the continuum. *Phys. Rev. Lett.* **2012**, *108*, 070401. [CrossRef]
16. Song, Y.; Jiang, N.; Liu, L.; Hu, X.; Zi, J. Cherenkov radiation from photonic bound states in the continuum: Towards compact free-electron lasers. *Phys. Rev. Appl.* **2018**, *10*, 064026. [CrossRef]
17. Kodigala, A.; Lepetit, T.; Gu, Q.; Bahari, B.; Fainman, Y.; Kanté, B. Lasing action from photonic bound states in continuum. *Nature* **2017**, *541*, 196–199. [CrossRef]
18. Zhang, H.; Wang, T.; Tian, J.; Sun, J.; Li, S.; De Leon, I.; Zaccaria, R.P.; Peng, L.; Gao, F.; Lin, X. Quasi-BIC laser enabled by high-contrast grating resonator for gas detection. *Nanophotonics* **2021**, *11*, 297–304. [CrossRef]
19. Hwang, M.-S.; Lee, H.-C.; Kim, K.-H.; Jeong, K.-Y.; Kwon, S.-H.; Koshelev, K.; Kivshar, Y.; Park, H.-G. Ultralow-threshold laser using super-bound states in the continuum. *Nat. Commun.* **2021**, *12*, 4135. [CrossRef]
20. Romano, S.; Lamberti, A.; Masullo, M.; Penzo, E.; Cabrini, S.; Rendina, I.; Mocella, V. Optical biosensors based on photonic crystals supporting bound states in the continuum. *Materials* **2018**, *11*, 526. [CrossRef]
21. Romano, S.; Zito, G.; Torino, S.; Calafiore, G.; Penzo, E.; Coppola, G.; Cabrini, S.; Rendina, I.; Mocella, V. Label-free sensing of ultralow-weight molecules with all-dielectric metasurfaces supporting bound states in the continuum. *Photonics Res.* **2018**, *6*, 726–733. [CrossRef]
22. Li, Z.; Xiang, Y.; Xu, S.; Dai, X. Ultrasensitive terahertz sensing in all-dielectric asymmetric metasurfaces based on quasi-BIC. *JOSA B* **2022**, *39*, 286–291. [CrossRef]

23. Azzam, S.I.; Shalaev, V.M.; Boltasseva, A.; Kildishev, A.V. Formation of bound states in the continuum in hybrid plasmonic-photonic systems. *Phys. Rev. Lett.* **2018**, *121*, 253901. [CrossRef] [PubMed]
24. Cao, J.; Sun, Y.; Kong, Y.; Qian, W. The sensitivity of grating-based SPR sensors with wavelength interrogation. *Sensors* **2019**, *19*, 405. [CrossRef] [PubMed]
25. West, P.R.; Ishii, S.; Naik, G.V.; Emani, N.K.; Shalaev, V.M.; Boltasseva, A. Searching for better plasmonic materials. *Laser Photonics Rev.* **2010**, *4*, 795–808. [CrossRef]
26. Johnson, P.B.; Christy, R.-W. Optical constants of the noble metals. *Phys. Rev. B* **1972**, *6*, 4370. [CrossRef]
27. Pastrňák, J.; Roskovcová, L. Refraction index measurements on AlN single crystals. *Phys. Status Solidi B* **1966**, *14*, K5–K8. [CrossRef]
28. Yeh, Y.-L. Real-time measurement of glucose concentration and average refractive index using a laser interferometer. *Opt. Lasers Eng.* **2008**, *46*, 666–670. [CrossRef]
29. Iqbal, T.; Khalil, S.; Ijaz, M.; Riaz, K.N.; Khan, M.I.; Shakil, M.; Nabi, A.G.; Javaid, M.; Abrar, M.; Afsheen, S. Optimization of 1D plasmonic grating of nanostructured devices for the investigation of plasmonic bandgap. *Plasmonics* **2019**, *14*, 775–783. [CrossRef]
30. Blaber, M.G.; Arnold, M.D.; Ford, M.J. A review of the optical properties of alloys and intermetallics for plasmonics. *J. Phys. Condens. Matter* **2010**, *22*, 143201. [CrossRef]
31. Sharma, A.K.; Pandey, A.K. Metal oxide grating based plasmonic refractive index sensor with Si layer in optical communication band. *IEEE Sens. J.* **2019**, *20*, 1275–1282. [CrossRef]
32. Lu, X.; Zheng, G.; Zhou, P. High performance refractive index sensor with stacked two-layer resonant waveguide gratings. *Results Phys.* **2019**, *12*, 759–765. [CrossRef]
33. Chou Chau, Y.-F.; Chou Chao, C.-T.; Huang, H.J.; Kooh, M.R.R.; Kumara, N.; Lim, C.M.; Chiang, H.-P. Perfect dual-band absorber based on plasmonic effect with the cross-hair/nanorod combination. *Nanomaterials* **2020**, *10*, 493. [CrossRef] [PubMed]
34. Sun, P.; Zhou, C.; Jia, W.; Wang, J.; Xiang, C.; Xie, Y.; Zhao, D. Narrowband absorber based on magnetic dipole resonances in two-dimensional metal–dielectric grating for sensing. *Opt. Commun.* **2020**, *459*, 124946. [CrossRef]
35. Li, Y.; Liu, Y.; Liu, Z.; Tang, Q.; Shi, L.; Chen, Q.; Du, G.; Wu, B.; Liu, G.; Li, L. Grating-assisted ultra-narrow multispectral plasmonic resonances for sensing application. *Appl. Phys. Express* **2019**, *12*, 072002. [CrossRef]
36. Sreekanth, K.V.; Alapan, Y.; ElKabbash, M.; Ilker, E.; Hinczewski, M.; Gurkan, U.A.; De Luca, A.; Strangi, G. Extreme sensitivity biosensing platform based on hyperbolic metamaterials. *Nat. Mater.* **2016**, *15*, 621–627. [CrossRef]
37. He, W.; Feng, Y.; Hu, Z.-D.; Balmakou, A.; Khakhomov, S.; Deng, Q.; Wang, J. Sensors with multifold nanorod metasurfaces array based on hyperbolic metamaterials. *IEEE Sens. J.* **2019**, *20*, 1801–1806. [CrossRef]

Article

Wavelength-Tunable Optical Two-Tone Signals Generated Using Single Mach-Zehnder Optical Modulator in Single Polarization-Mode Sagnac Interferometer

Akito Chiba [1,*] **and Yosuke Akamatsu** [2]

[1] Faculty of Science and Technology, Gunma University, Kiryu-shi 376-8515, Gunma, Japan
[2] Graduate School of Science and Technology, Gunma University, Kiryu-shi 376-8515, Gunma, Japan; t10306001@gunma-u.ac.jp
* Correspondence: akito.chiba@osamember.org or chiba@gunma-u.ac.jp

Abstract: We demonstrate 60 GHz separation optical two-tone signal generation at arbitrary C-band wavelengths without involving complicated optical wavelength filtering. By utilizing a polarizer, the selective suppression of undesired low-order optical sidebands has been proven and optimized based on model analysis. By utilizing this scheme in conjunction with the optimized parameters, more than 20 dB of suppression of undesired optical sidebands have been successfully achieved over a 40 nm wavelength range. This scheme allows us to generate optical two-tone signals at the desired wavelength.

Keywords: microwave photonics; optical modulation; optical polarization; optical two-tone signals; RF photonics

Citation: Chiba, A.; Akamatsu, Y. Wavelength-Tunable Optical Two-Tone Signals Generated Using Single Mach-Zehnder Optical Modulator in Single Polarization-Mode Sagnac Interferometer . *Photonics* **2022**, *9*, 194. https://doi.org/10.3390/photonics9030194

Received: 20 January 2022
Accepted: 8 March 2022
Published: 17 March 2022

1. Introduction

A phase-synchronized pair of monochromatic lightwaves with stable frequency spacing plays an important role in the complementary use of radio-wave (RF) signals and lightwaves, i.e., in the field of microwave photonics. Such a lightwave-pair is called as an optical two-tone (OTT) signal. Since the RF signal obtained from the direct detection of the OTT signal has a stable frequency, this can be applied to the convergence of optical and wireless communications [1], high-resolution-image broadcasting [2], precise clock distribution [3,4], radar measurement [5], and THz signal generation [6]. Not limited in the complementary use of an RF signal and lightwaves, some advantages are there in RF frequency upconversion, such as low phase noise, frequency tunability, increase in the output signal frequency and allowance of some optical techniques such as optical amplification.

Wide frequency-separation OTT signals are especially useful because of the demand for millimeter-waveband RF signals; therefore, several types of OTT signal generation schemes have been explored. One involves constructing a phase-locked loop for lightwaves [7]. Its offset range currently reaches 17.8 GHz [8] due to phase noise (linewidth) suppression of the semiconductor lasers constituting the optical phase-locked loop. However, a significant problem remains in further increasing the bandwidth: it requires broad loop-BW, which implies that the closed-loop should be as short as possible. In contrast, there is another approach that employs optical modulation to generate optical sidebands [9], which is facilitated by the development of a waveguide-type optical modulator equipped with traveling-wave electrodes [10,11]. In this scheme, the optical frequency spacing of the generated OTT signals is mainly dominated by the driving frequency of the optical modulator, and it can be exceeded by extracting a pair of higher-order optical sidebands generated by deep optical modulation.

Several approaches have been demonstrated for sideband extraction. The straightforward way is to filter an optical signal in the wavelength domain [12–14], which involves the precise adjustment of transmission/rejection wavelength ranges when the wavelength of the OTT signals should be changed. Additionally, their optical-frequency separation is limited by the steepness of the optical filter in the wavelength axis. The other approaches are investigated using a modulation order-dependent interference by utilizing an I-Q optical modulator [15–17] and periodic phase shifts of each optical sideband given by RF signals; the desired optical sidebands survive to become OTT signals via constructive optical interference, while unnecessary optical sidebands disappear due to the destructive optical interference [18–20]. The approach is an optical filter-free operation, allowing for the optical wavelength flexibility to endure. However, the optical phase offset must be specifically stabilized to achieve low-spurs operation in long term, and an optical modulation device should be densely integrated to increase the degree of frequency multiplication (i.e., the ratio of frequency spacing of OTT signals against the frequency of the RF signal driving the optical modulator). Although densely-integrated optical modulation devices have been reported [21–25], further dense integration would be limited via a fabrication process and physical parameters such as wavelength of the lightwave and refractive indices of materials.

Another approach involves using the polarization of light as one degree of freedom. While some attempts were conducted using a polarization modulator in a polarization-maintaining Sagnac interferometer (PMSI) [26,27], one of the issues was stability degradation. In these approaches, BOTH polarization modes of PMSI were utilized, so that the output signal would be unstable due to a fluctuation in polarization by temperature via retardation in optical fibre. Deviation of the bias voltage of the polarization modulator would also induce degradation of its performance. A system for stabilizing static optical phases must be required, as well as the approaches based on an integrated optical modulator. Furthermore, the analytically obtained optimization results were complicated.

These problems have been solved using a configuration where a bi-directional single Mach-Zehnder optical modulator (MZM) is nested in only ONE polarization mode of the modified PMSI [28–31]. In this article, we describe how this approach allows us to generate OTT signals at arbitrary wavelengths. Our scheme promises wavelength tunability and simplicity; it only requires wavelength changes for the seed lightwave of the OTT signal generator, and it is free from the precise adjustment of an optical band-rejection filter. In Section 2, the operation principle is described by introducing an analysis model to derive an equation of an output lightwave signal from the proposed configuration. The analytical results are also discussed. In Section 3, we describe a proof-of-principle experiment. Based on the obtained optical spectra and their derivatives, we show that this scheme is suitable to generate the OTT signals for C-band wavelengths. In Section 4, we summarize the OTT signal generation scheme.

2. Principle

2.1. Output Signal from Polarization-Maintaining Sagnac Interferometer

Figure 1 shows a model for generating wavelength-tunable optical two-tone signals, which is composed of a push-pull-driven MZM within one mode of a PMSI [30,31]. In this setup, P-polarized incident lightwaves propagate in the clockwise direction, while S-polarized ones propagate in the counter-clockwise direction. Due to the polarization-rotation element (PRE), both components become S (TE) polarized at the MZM. Incident lightwave $\boldsymbol{E}_0$, composed of the P-polarization component $E_{0\mathrm{P}}$ and S-polarization component $E_{0\mathrm{S}}$, is described as

$$\boldsymbol{E}_0 = \begin{bmatrix} E_{0\mathrm{S}} \\ E_{0\mathrm{P}} \end{bmatrix} = \begin{bmatrix} E_0 \cos\alpha \\ E_0 \sin\alpha \end{bmatrix} \tag{1}$$

where α is the angle of the polarizer placed at the input port of polarizing beam splitter (PBS), and E_0 is the lightwave amplitude just after passing through the polarizer. Hereafter, the polarization-extinction ratio (i.e., the inverse of attenuation of the polarization component when the insertion loss of the polarizer is omitted) is denoted as ξ, while in Equation (1) the effect of the polarization extinction ratio is omitted because the dominant term of each polarization component belongs to its polarization axis. The lightwaves after passing through the MZM and PRE become

$$E_{CCW} = \begin{bmatrix} 0 \\ -T_{MZM}(\Delta\theta,\ \theta_B)E_{0S} \end{bmatrix} = \begin{bmatrix} 0 \\ -T_{MZM}(\Delta\theta\ ,\ \theta_B)E_0 \cos\alpha \end{bmatrix} \quad (2)$$

and

$$E_{CW} = \begin{bmatrix} T_{MZM}(\eta\Delta\theta,\ \theta_B)E_{0P} \\ 0 \end{bmatrix} = \begin{bmatrix} T_{MZM}(\eta\Delta\theta,\ \theta_B)E_0 \sin\alpha \\ 0 \end{bmatrix} \quad (3)$$

for counter-clockwise propagation E_{CCW} and clockwise propagation E_{CW}, respectively. Here, we assume that the polarization extinction ratio of the PBS is infinite; however, the effect of its finite polarization extinction ratio results in lightwave power loss due to leakage at the empty port of the PBS in the PMSI. Note that, at the PRE, the P-polarized lightwave is converted into an S-polarized one, and vice versa. T_{MZM} $(\Delta\theta, \theta_B)$ is the transmittance of the MZM that is driven by a sinusoidal RF signal with an angular frequency of ω_0 and induced optical phase $\Delta\theta$, which is given by

$$T_{MZM}(\Delta\theta,\ \theta_B) = \cos(\Delta\theta \sin\omega_0 t + \theta_B), \quad (4)$$

if the insertion loss of the MZM is omitted. θ_B is the phase bias of the MZM, and hereafter θ_B is assumed to be $\pi/2$; i.e., the MZM is driven under the null-bias condition. η is the ratio between the two induced optical phases, and $|\eta|$ is assumed to be less than 1. So, the attenuation of the RF signal amplitude to achieve the ratio is $1/|\eta|$. In Figure 1, a set of induced optical phases ($\Delta\theta$ and $\eta\Delta\theta$) is independently shown, but actually, the set can beprepared by utilizing RF signal reflection or circulation. Adopting the MZM possessing RF termination ports, we can simultaneously induce two optical modulations with different modulation indices [29]. Using this feature, we modulate the two circulating lightwaves independently, with modulation indices of $\Delta\theta$ and $\eta\Delta\theta$ for E_{CCW} and E_{CW}, respectively.

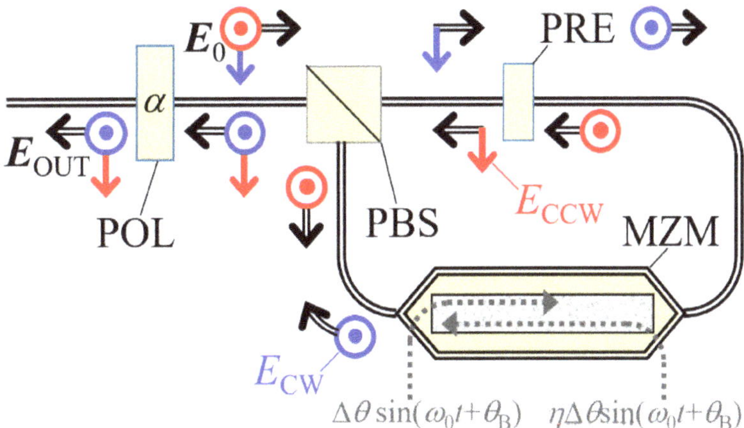

Figure 1. Analysis model of wavelength-tunable optical two-tone signals generation based on an MZM in PMSI with PRE. PBS: polarizing beam splitter, PRE: polarization-rotation element, POL: polarizer. Blue solid arrows and red solid arrows depict the electric field (polarization) of lightwave propagating in the PMSI in the clockwise direction and the counter-clockwise direction, respectively. In the configuration, lightwaves propagate according to black open arrows. Dashed arrows indicate the RF signal modulating lightwave in the PMSI.

At PBS, these lightwaves are combined and projected onto the polarizer. The lightwave, after passing through the polarizer, E_{OUT}, can be described as

$$
\begin{aligned}
\boldsymbol{E}_{\mathrm{OUT}} &= \boldsymbol{T}_{\mathrm{POL}}[\boldsymbol{E}_{\mathrm{CCW}} + \boldsymbol{E}_{\mathrm{CW}}] \\
&= \begin{bmatrix} \cos\alpha & \sin\alpha \\ -\sin\alpha & \cos\alpha \end{bmatrix} \begin{bmatrix} 1 & 0 \\ 0 & 1/\sqrt{\xi} \end{bmatrix} \begin{bmatrix} \cos\alpha & -\sin\alpha \\ \sin\alpha & \cos\alpha \end{bmatrix} \times \begin{bmatrix} T_{\mathrm{MZM}}(\eta\Delta\theta,\, \theta_B)E_{0\mathrm{P}} \\ -T_{\mathrm{MZM}}(\Delta\theta,\, \theta_B)E_{0\mathrm{S}} \end{bmatrix} \\
&= \tfrac{E_0 \sin 2\alpha}{2}\{T_{\mathrm{MZM}}(\Delta\theta,\, \theta_B) + T_{\mathrm{MZM}}(\eta\Delta\theta,\, \theta_B)\} \times \begin{bmatrix} \cos\alpha \\ -\sin\alpha \end{bmatrix} \\
&+ \tfrac{E_0}{\sqrt{\xi}}\{T_{\mathrm{MZM}}(\eta\Delta\theta,\, \theta_B)\sin^2\alpha - T_{\mathrm{MZM}}(\Delta\theta,\, \theta_B)\cos^2\alpha\} \times \begin{bmatrix} \sin\alpha \\ \cos\alpha \end{bmatrix},
\end{aligned} \tag{5}
$$

where the first and second terms correspond to the lightwave amplitude parallel and perpendicular to the axis of the polarizer, respectively. $\boldsymbol{T}_{\mathrm{POL}}$ is the amplitude transmittance of the polarizer:

$$
\boldsymbol{T}_{\mathrm{POL}} = \begin{bmatrix} \cos\alpha & \sin\alpha \\ -\sin\alpha & \cos\alpha \end{bmatrix} \begin{bmatrix} 1 & 0 \\ 0 & 1/\sqrt{\xi} \end{bmatrix} \begin{bmatrix} \cos\alpha & -\sin\alpha \\ \sin\alpha & \cos\alpha \end{bmatrix}. \tag{6}
$$

If the orthogonal axis of $\boldsymbol{E}_{\mathrm{OUT}}$ is chosen to be parallel and perpendicular to the polarizer, the lightwave $\boldsymbol{E}'_{\mathrm{OUT}}$ can be expressed as

$$
\begin{aligned}
\boldsymbol{E}'_{\mathrm{OUT}} &= \boldsymbol{T}'_{\mathrm{POL}}[\boldsymbol{E}_{\mathrm{CCW}} + \boldsymbol{E}_{\mathrm{CW}}] \\
&= \begin{bmatrix} 1 & 0 \\ 0 & 1/\sqrt{\xi} \end{bmatrix} \begin{bmatrix} \cos\alpha & -\sin\alpha \\ \sin\alpha & \cos\alpha \end{bmatrix} \begin{bmatrix} T_{\mathrm{MZM}}(\eta\Delta\theta,\, \theta_B)E_{0\mathrm{P}} \\ -T_{\mathrm{MZM}}(\Delta\theta,\, \theta_B)E_{0\mathrm{S}} \end{bmatrix} \\
&= E_0 \begin{bmatrix} \tfrac{\sin 2\alpha}{2}\{T_{\mathrm{MZM}}(\Delta\theta,\, \theta_B) + T_{\mathrm{MZM}}(\eta\Delta\theta,\, \theta_B)\} \\ \tfrac{-1}{\sqrt{\xi}}\{T_{\mathrm{MZM}}(\Delta\theta,\, \theta_B)\cos^2\alpha - T_{\mathrm{MZM}}(\eta\Delta\theta,\, \theta_B)\sin^2\alpha\} \end{bmatrix}.
\end{aligned} \tag{7}
$$

The lightwave coming back to the polarizer from the PMSI is projected with a different angle, $-\alpha$.

2.2. Suppression of First-Order Sideband for Third-Order Sideband Extraction

$\boldsymbol{E}_{\mathrm{out}}$ is composed of many optical-frequency components when MZM is driven by strong sinusoidal RF signals. However, assuming that infinite ξ, Equations (5) and (7) imply that some optical-frequency components of $\boldsymbol{E}_{\mathrm{out}}$ become zero under a certain condition. This fact can be derived by expanding the term $T_{\mathrm{MZM}}(\Delta\theta, \theta_B) + T_{\mathrm{MZM}}(\eta\Delta\theta, \theta_B)$ using the m-th order Bessel function of the first kind $J_m(x)$:

$$
\begin{aligned}
& T_{\mathrm{MZM}}(\Delta\theta,\, \theta_B) + T_{\mathrm{MZM}}(\eta\Delta\theta,\, \theta_B) \\
&= +j \sum_{m=-\infty}^{\infty} [J_{2m+1}(\Delta\theta) + J_{2m+1}(\eta\Delta\theta)]\, e^{j(2m+1)\omega_0 t}
\end{aligned} \tag{8}
$$

Then, the $(2m + 1)$-th order optical sidebands disappear when the following equation is satisfied:

$$
J_{2m+1}(\Delta\theta) + J_{2m+1}(\eta\Delta\theta) = 0. \tag{9}
$$

From the above equation, adequate parameters can be obtained for suppressing ±1st order optical sidebands (i.e., m = –1, 0) in the lightwave composed of –3rd~+3rd order optical sidebands. Note that, ideally, the null-biased MZM does not generate even-order optical sidebands. Although there are many solutions satisfying Equation (9), we focus on those with a negative η which can be achieved by a π-phase shift of the RF signal modulating clockwise lightwave [30,31]; i.e., η is in the range from –1 to 0. Additionally, η = –1 is not suitable, because Equation (9) is satisfied for arbitrary $\Delta\theta$ and m, meaning that all sidebands disappear. Figure 2a shows the plot of $J_1(x)$ versus x to find a numerical solution of Equation (9), and the pairs of the solution are shown in Figure 2b.

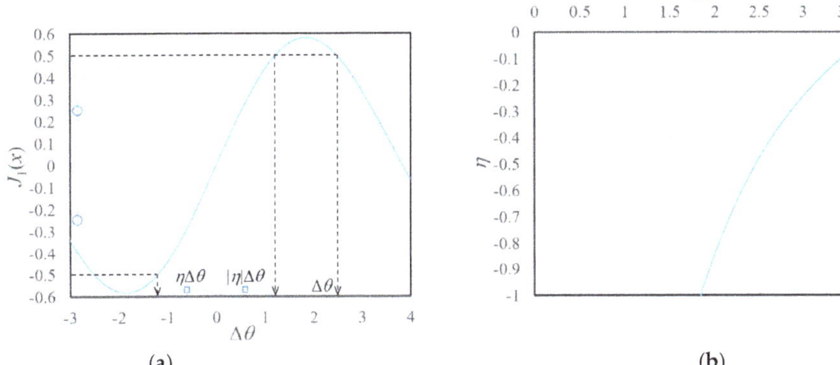

Figure 2. (**a**) A solution of Equation (9) for $m = 0$ and $\Delta\theta = 2.485$, and (**b**) the combination of η and $\Delta\theta$ satisfying Equation (9).

Adapting the parameter setting, the polarization of input ($\alpha = 45°$) and output lightwaves can be drawn as Figure 3. While polarization of the lightwave launching into the PMSI is parallel to the axis of the polarizer (Figure 3a), $\pm$1st-order optical sidebands of $\boldsymbol{E}_{\mathrm{CW}}$ (originating from the P-polarization incident lightwave) are flipped in their phase at a weak optical modulation while those of $\boldsymbol{E}_{\mathrm{CCW}}$ retains their phase. Then, as shown in Figure 3b, the polarization of $\pm$1st-order optical sidebands is perpendicular to the polarization axis of the polarizer. In contrast, $\pm$3rd-order optical sidebands are sufficiently small for the S-polarization components due to the modulation index including a small $|\eta|$; hence, only the sidebands of the P-polarization are projected by the polarizer so that they become the desired output lightwave, as shown in Figure 3c. Under a negative η, the intensity of the desired $\pm$3rd-order optical sidebands, P_3, is approximately expressed as

$$P_3 \simeq E_0^2 \frac{\sin^2 2\alpha}{4} [J_3(\Delta\theta) - J_3(|\eta|\Delta\theta)]^2, \tag{10}$$

when the components perpendicular to the axis of the polarizer are neglected. For the $\pm$1st-order optical sidebands, which are undesired components in the generated OTT signals, P_1 is

$$\begin{aligned} P_1 \simeq\ & E_0^2 \Big[\tfrac{\sin^2 2\alpha}{4} [J_1(\Delta\theta) - J_1(|\eta|\Delta\theta)]^2 \\ & + \left(\tfrac{\cos^2 \alpha J_1(\Delta\theta) + \sin^2 \alpha J_1(|\eta|\Delta\theta)}{\sqrt{\xi}} \right)^2 \Big]. \end{aligned} \tag{11}$$

From Equations (10) and (11), the optimum value of α is expected to be 45°. Figure 4a shows the dependence of P_1/P_3 on $\Delta\theta$ for several $|\eta|$, assuming that $\xi = 10^4$ (40 dB). The $\Delta\theta$, giving the bottom of each dip, corresponds to the solution of Equation (9). With a decreasing $|\eta|$ (i.e., an increase in attenuation), the dip of each plot gradually shifts to the higher $\Delta\theta$. Additionally, the dip becomes narrower with a decrease in $|\eta|$, originating from the fact that the slope of $J_1(x)$ increases when the argument x is far from $x_0 = 1.841$, which gives a local maximum of $J_1(x_0)$. This fact is also summarized in Figure 4b, which show dependence of the maximum suppression ratio of the 1st-order optical sidebands and 3 dB width of the dips versus $|\eta|$. This means that, by adopting a lower $|\eta|$ (higher attenuation), the suppression ratio is enhanced by more than 30 dB, while the suppression ratio is degraded by deviations of η and $\Delta\theta$. A deviation in α has less of an effect on the suppression ratio given a sufficiently high polarization extinction ratio, while the deviation directly affects P_3 and P_1, as implied by Equations (10) and (11).

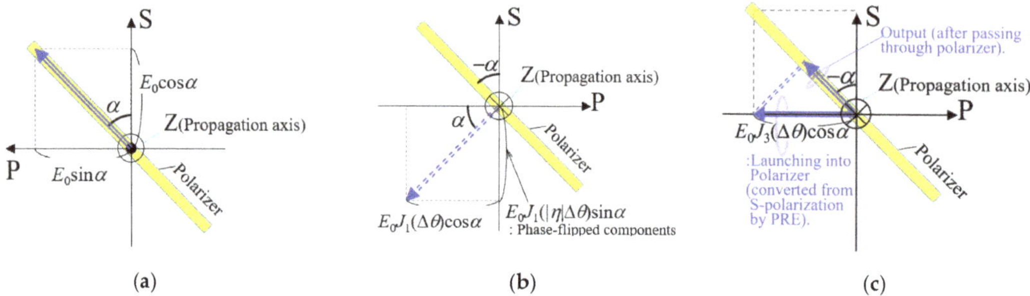

Figure 3. Alignment of the polarization axis of the polarizer and polarization state of an (**a**) incident lightwave and the output lightwave composed of (**b**) ±1st-order optical sidebands and (**c**) ±3rd-order optical sidebands. Note that the propagation axis (Z) of (**a**) is opposite to that of (**b**,**c**), while the spatial coordinate is consistent.

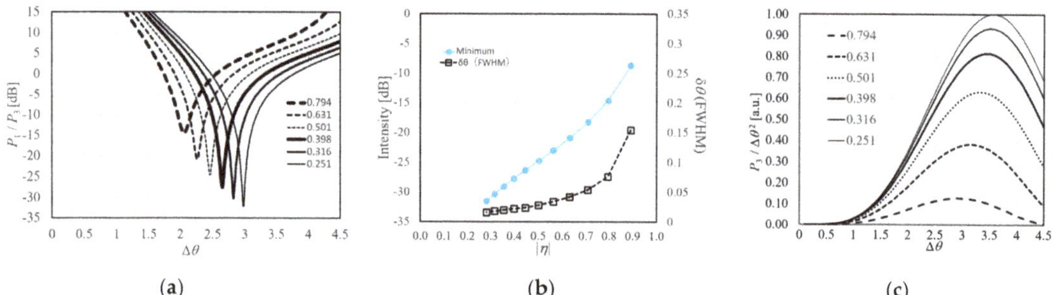

Figure 4. (**a**) Dependence of 1st-order optical sideband intensity on $\Delta\theta$ when η is in negative and its absolute value $|\eta|$ equals to 0.794 (2 dB attenuation), 0.631 (4 dB attenuation), 0.501 (6 dB attenuation), 0.398 (8 dB attenuation), 0.316 (10 dB attenuation), 0.251 (12 dB attenuation). Polarization extinction ratio ξ is assumed to be 10^4. All plots are normalized by the intensity of ±3rd-order optical sidebands obtained from each η. (**b**) Minimum intensity and 3-dB width of the dips in (**a**) versus $|\eta|$, ranging from 1 dB to 11 dB with a step of 1 dB in RF attenuation $1/|\eta|$. (**c**) Output power of desired wavelength component P_3 normalized by $\Delta\theta^2$ being proportional to input RF power. Calculation is performed using Equation (10) and setting of $|\eta|$ is common with (**a**). All traces are normalized by the maximum when $|\eta|$ equals to 0.251.

Regarding the desired optical signal power P_3, the output power is increased with increase of $\Delta\theta$, according to the characteristic of $J_3(x)$. For focusing on conversion efficiency from the input RF drive signal power, P_3 normalized by $\Delta\theta^2$ is plotted on Figure 4c for several $|\eta|$. As can be seen, $P_3/\Delta\theta^2$ is increased with a decreasing $|\eta|$ and the $\Delta\theta$ giving the local maxima of each plot are gradually increased, in the range of 3–3.5 in $\Delta\theta$.

3. Experiments

3.1. Experimental Setup and Proof-of-Concept Experiment

For conducting an experiment, we evaluate the wavelength dependence of the extinction ratio (ER) of the MZM integrated into a Z-cut Lithium Niobate substrate [31]. Halfwave voltage and modulation bandwidth of the modulation electrodes were evaluated to be 2.4 V and 23 GHz respectively, for each arm. The ER was designed to be more than 20 dB, and the insertion loss was evaluated to be 5.7 dB, at a wavelength of 1550 nm. To evaluate the wavelength dependence of ER, optical-power transmission spectra of MZM were obtained using a wavelength-swept light source (Agilent, 81689A) with a line-width of around 1 MHz. Transmission spectra of MZM were measured under each of its null- and in-phase conditions at a wavelength of 1550 nm, and the condition was fixed during each of the spectrum measurements. For wavelengths ranging from 1530 nm to 1570 nm, ERs of more than 30 dB were obtained from the MZM. And, at the wavelength of 1550 nm, ER was evaluated to be more than 40 dB. Then, the suppression of undesired even-order sidebands and carriers is sufficiently guaranteed due to destructive interference within MZM.

Figure 5 shows the experimental setup for evaluating the wavelength tunability of the OTT signal generator [30,31]. For the lightsource of the OTT signal generator, we employed an external-cavity laser diode (Agilent, 81689A). By using a 2 × 2 optical coupler followed by a polarization-maintaining optical circulator (OC), a polarizer (polarization extinction ratio >35 dB) and a quarter-waveplate, the incident lightwave generated from the lightsource was introduced into PMSI, which was composed of a polarizing beam splitter (polarization extinction ratio >21 dB, insertion loss <1.0 dB), PRE and MZM. To compensate for the wavelength-dependence of polarization-mode dispersion of polarization-maintaining optical fibers (PMFs), some PMFs were connected to couple its slow- (fast-) axis to the fast- (slow-) axis of the other PMFs. In addition, a quarter waveplate was employed to rotate the polarization of parasitic unmodulated lightwaves originating from the reflection at the end of a PMF and/or PBS; hence, the parasitic unmodulated lightwaves were rejected by the polarizer. Among the setup, the components with a narrow wavelength range were the OC (1550 ± 30 nm) and the 2 × 2 optical coupler (1550 ± 20 nm), which restricts the tunable range of the setup. However, the latter was used just for monitoring the launched optical power, so that it can be removed from the setup. The bias voltage of the MZM was adjusted to its null-bias condition at the wavelength of 1550 nm to suppress the optical carrier and even-order optical sidebands using the MZM. During measurement, the bias voltages were fixed. To drive the MZM, a 10 GHz RF signal was amplified (Ciao Wireless, CA-910-4042) and applied to the MZM. The applied RF power was evaluated using a conventional RF power meter (HP, 437B and 8481A). Using an optical spectrum analyzer, the optical spectrum of the lightwave emitted from the OTT signal generator was evaluated.

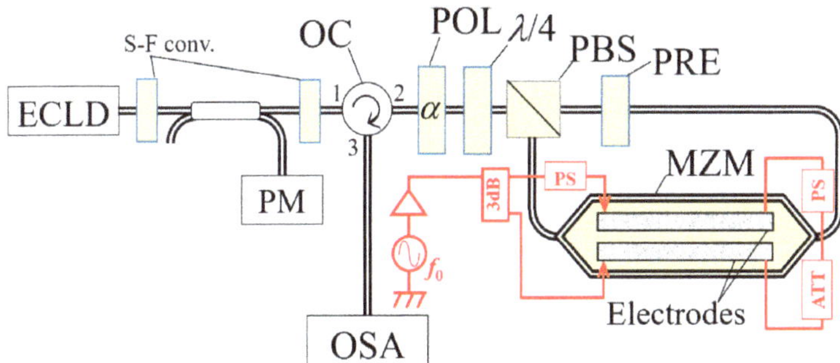

Figure 5. Experimental setup. Thin red lines are the RF signal connections, and double solid lines are connections for lightwave. ECLD: External-cavity diode laser, S-F conv.: PMF connector coupling a slow- (fast-) axis of the either PMF to a fast- (slow-) axis of the other, PM: optical power-meter, OC: Optical circulator, POL: Polarizer, $\lambda/4$: quarter-waveplate, PBS: Polarizing beam-splitter, PRE: Polarization-rotation element, MZM: Mach-Zehnder optical Modulator, OSA: optical spectrum analyzer, 3 dB: 3 dB RF hybrid coupler, PS: RF phase shifter, and ATT: Attenuator.

3.2. Experimental Results Obtained from 1550 nm Seed Lightwave

Figure 6 shows the typical optical spectra of the experimentally obtained OTT signals and the strongly modulated lightwave launching into PBS. The RF signal amplitude was adjusted to set the modulation index $\Delta\theta$ to 2.93, and the RF attenuation $1/|\eta|$ required for selective polarization rotation was adjusted to 11 dB, i.e., two modulation indices $\Delta\theta$ and $|\eta|\Delta\theta$ (=0.83) were obtained from the single MZM. As can be seen, the ±1st-order optical sidebands were suppressed by 34 dB by the polarizer, and the desired ±3rd-order optical sidebands were −28 dBm to −29 dBm. Although no feedback system was adopted in the experimental setup simply placed in a room without strict thermal shielding, the obtained optical output power was sufficiently stable; power fluctuation of the suppressed ±1st-order optical sidebands was within ±0.3 dB for 8 h. This implies that the phase difference between the lightwaves circulating in PMSI does not fluctuate due to the use of the same one polarization mode of PMSI. Using both polarization modes in the PMSI composed of PMFs, unavoidable fluctuation in optical-phase difference would be induced by temperature fluctuation. It should be noted that only the ±1st-order optical sidebands to be suppressed at the polarizer are involved in the effect of the phase-difference fluctuation, while the other sidebands and carrier do not suffer from such an effect: although intensity fluctuation of even-order ones including the carrier might originate from that of MZM bias, and its degree is sufficiently small. Additionally, the other odd-order (in this case ±3rd- and ±5th-order) ones do not undergo interference so that their intensity is also stable.

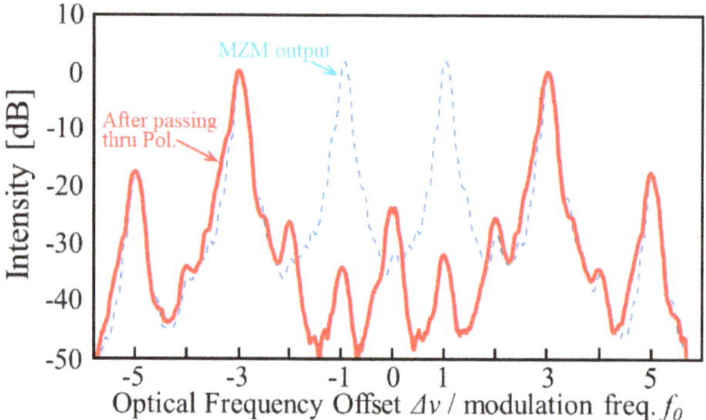

Figure 6. Optical spectrum of 60-GHz separation OTT signal (Solid red line) composed of strongly modulated lightwave (Dashed blue line) and weakly modulated lightwave. The frequency of RF signal driving MZM, f_0, is 10 GHz. Horizontal and vertical axes are normalized by f_0 and the intensity of ±3rd-order sidebands, respectively.

To evaluate the suppression of ± 1st-order optical sidebands, we experimentally evaluate the dependence of the intensity on the induced optical phase $\Delta\theta$ under the fixed η, as shown in Figure 7. As can be seen, the intensity plots agree well with the analytically obtained dip characteristic for RF attenuation $1/\mid\eta\mid$ = 11.5 dB, supporting the validity of the model analysis shown in Figure 4a of Section 2.2. The 0.5-dB difference against the RF attenuation in the setup (11 dB) may be due to the accumulation of the other insertion losses such as DC blocks and bias tees, and the residual calibration error of the RF power meter. The experimentally obtained intensity was −35 dB, which was restricted by the lower limit of the optical spectrum analyzer used in the experiment and the ER of the polarizer. The degree of the suppressed intensity is also evaluated under the condition of satisfying Equation (9) for the negative η, which is shown in Figure 8. With a decrease in $\Delta\theta$, the intensity gradually decreased and followed the analytical result. Some deviation in the experimental results would be mainly due to the deviations of RF power and RF attenuator, besides instrument accuracy of the optical spectrum analyzer.

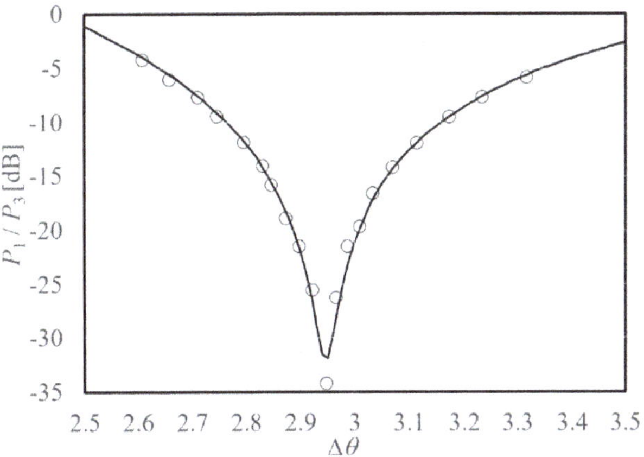

Figure 7. Intensity of ±1st-order optical sidebands obtained from the experiment (open circles) and analytical model when RF attenuation $1/|\eta|$ =11.5 dB (solid line), normalized by ±3rd-order optical sideband intensity.

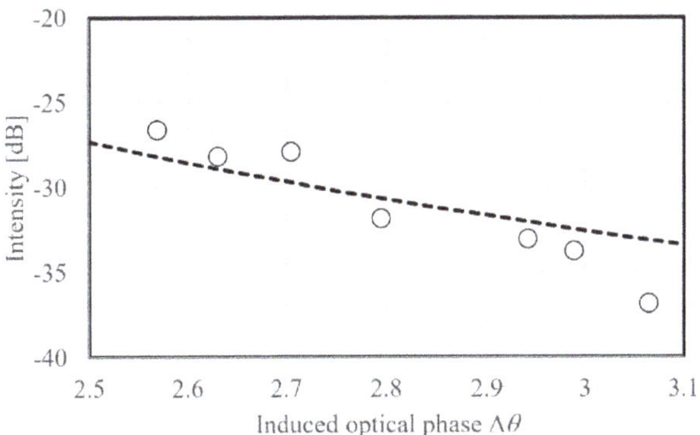

Figure 8. Intensity of ±1st-order optical sidebands normalized by ±3rd-order optical sideband intensity, obtained from the experiment (open circles) and analytical model where ER of the polarizer is 41.5 dB (dashed line). At each $\Delta\theta$, η is set to satisfy Equation (9).

Figure 9 shows RF spurs against $\Delta\theta$ obtained from a photodiode (PD) with a 3 dB bandwidth of 70 GHz (Finisar, XPDV3120R) and an RF harmonic mixer (HP, 11970U) connected to an RF spectrum analyzer (HP, 8563E). First- (10 GHz) and fourth-order (40 GHz) spurs were detected with the degree of ~−30 dB, mainly originating from the beat of the desired components, their neighbor components (±2nd-order components), and the ±1st-order components cut by the polarizer. In contrast, second-order (20 GHz) spurs were relatively strong: around −10 dB against the desired lightwave components. These spurs are due to the desired components and the ±5th-order components. Additionally, these spurs were stronger than the expected power. Such a difference might be ascribed to the effect of dispersion of optical fibers. However, the strength of these spurs might not be significant: these RF frequencies are sufficiently far from the desired sextupled signal frequency so that they can be rejected by RF signal processing such as using an RF band-pass filter [29].

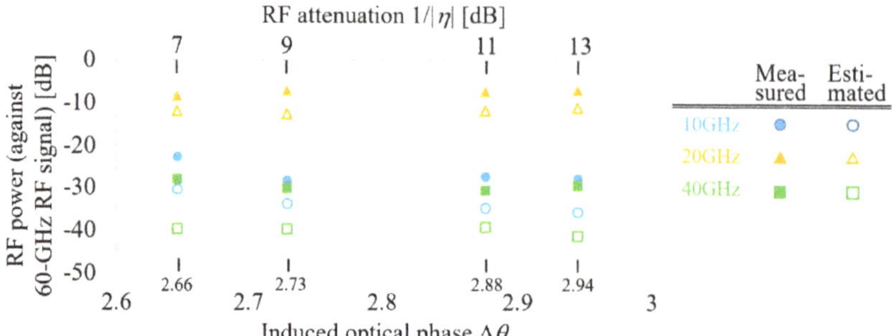

Figure 9. Power ratio of spurs versus the induced optical phase $\Delta\theta$. Origin of the vertical axis corresponds to 60-GHz RF power. Circles, triangles, and squares denote the 10 GHz (1st-order), 20 GHz (2nd-order), and 40 GHz (4th-order) RF signals. The filled symbols indicate experimental results, while the opened symbols indicate results from numerical calculations.

3.3. *Wavelength-Tunable OTT Signal Generation*

Figure 10 shows the typical optical spectra of OTT signals obtained from the generator with several seed lightwave wavelengths. The OTT signals were successfully generated without increasing the spurs: for the intensity ratio of ±1st-order optical sidebands against the desired ±3rd-order ones, more than 20 dB was achieved over a 40 nm wavelength range in the C-band, owing to the wavelength-independent ER of the MZM and the sufficiently high polarization extinction ratio of the polarizer in the broad-bandwidth range. In acquisition of Figure 10, the bias voltage of MZM was fixed at its null point for the wavelength of 1550 nm, to avoid a complicated adjustment of the bias voltage. Such a constant voltage might also induces residual spurs due to a slight deviation in the bias voltage condition from the null, when the wavelength of the seed lightwave shifts from 1550 nm. The suppression ratio would be further enhanced if we accept bias voltage tracking according to the wavelength of the seed lightwave. Power fluctuations of the desired OTT signals due to the seed-wavelength changes were within −25~−32 dBm.

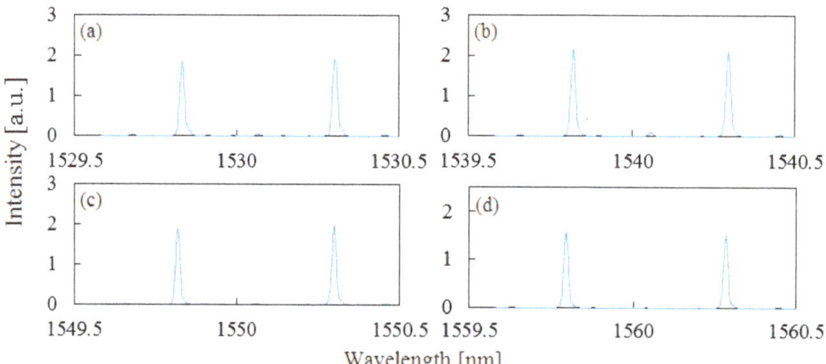

Figure 10. Optical spectra of 60-GHz separation OTT signals generated from the setup where the bias voltages of the MZM were fixed at its null point for the wavelength of 1550 nm. Wavelengths of seed lightwave were adjusted to (**a**) 1530.06 nm, (**b**) 1540.05 nm, (**c**) 1550.05 nm, and (**d**) 1560.04 nm, respectively.

4. Summary

We demonstrated an optical two-tone signal generation scheme at arbitrary wavelengths without involving complicated optical signal processing in the wavelength axis.

Using an analytical approach, we optimized the driving condition for suppressing undesired lightwave components and conducted a proof-of-principle experiment. Due to the sufficiently high ER of the MZM and polarizer, the $\pm$1st-order optical sidebands has been suppressed with a ratio of more than 20 dB against the desired wavelength components. Because this scheme requires no optical filters, it can be useful for generating optical two-tone signals at arbitrary wavelengths and may be extended to RF signal measurement by combining other microwave-photonics techniques.

Author Contributions: Conceptualization, A.C.; methodology, A.C.; software, Y.A. and A.C.; validation, A.C. and Y.A.; formal analysis, A.C. and Y.A.; investigation, Y.A. and A.C.; resources, A.C.; data curation, Y.A. and A.C.; writing—original draft preparation, A.C.; writing—review and editing, A.C. and Y.A.; visualization, A.C. and Y.A.; supervision, A.C.; project administration, A.C.; funding acquisition, A.C. All authors have read and agreed to the published version of the manuscript.

Funding: This work was supported in part by Japan Society for the Promotion of Science (20K21878, 20H02141) and the Telecommunications Advancement Foundation, Japan.

Institutional Review Board Statement: Not applicable.

Informed Consent Statement: Not applicable.

Data Availability Statement: Not applicable.

Acknowledgments: The authors would like to thank K. Higuma of Sumitomo Osaka Cement Co., Ltd. for supplying an MZM, and K. Takada, Y. Takahashi, and K. Miura for fruitful comments.

Conflicts of Interest: The authors declare no conflict of interest.

References

1. Shih, P.-T.; Lin, C.-T.; Jiang, W.-J.; Chen, Y.-H.; Chen, J.; Chi, S. Full duplex 60-GHz RoF link employing tandem single sideband modulation scheme and high spectral efficiency modulation format. *Opt. Express* **2009**, *17*, 19501–19508. [CrossRef] [PubMed]
2. Hirata, A.; Yamaguchi, R.; Kosugi, T.; Takahashi, H.; Murata, K.; Nagatsuma, T.; Kukutsu, N.; Kado, Y.; Iai, N.; Okabe, S.; et al. 10-Gbit/s Wireless Link Using InP HEMT MMICs for Generating 120-GHz-Band Millimeter-Wave Signal. *IEEE Trans. Microw. Theory Tech.* **2009**, *57*, 1102–1109. [CrossRef]
3. Qi, G.; Yo, J.; Seregelyi, J.; Paquet, S.; Belisle, C. Generation and distribution of a wide-band continuously tunable millimeter-wave signal with an optical external modulation technique. *IEEE Trans. Microw. Theory Tech.* **2005**, *53*, 3090–3097.
4. Kiuchi, H. Highly stable millimeter-wave signal distribution with an optical round-trip phase stabilizer. *IEEE Trans. Microw. Theory Tech.* **2008**, *56*, 1493–1500. [CrossRef]
5. Ghelfi, P.; Laghezza, F.; Scotti, F.; Serafino, G.; Capria, A.; Pinna, S.; Onori, D.; Porzo, C.; Scaffardi, M.; Malacarne, A.; et al. A fully photonics-based coherent radar system. *Nature* **2014**, *507*, 341–345. [CrossRef] [PubMed]
6. Nagatsuma, T.; Horiguchi, S.; Minamikata, Y.; Yoshimizu, Y.; Hisatake, S.; Kuwano, S.; Yoshimoto, N.; Terada, J.; Takahashi, H. Terahertz wireless communications based on photonics technologies. *Opt. Express* **2013**, *21*, 23736–23747. [CrossRef] [PubMed]
7. Armor, J.B., Jr.; Robinson, S.R. Phase-lock control considerations for coherently combined lasers. *Appl. Opt.* **1979**, *18*, 3165–3175. [CrossRef] [PubMed]
8. Simsek, A.; Arafin, S.; Kim, S.-K.; Morrison, G.B.; Johansson, L.A.; Mashanovitch, M.L.; Coldren, L.A.; Rodwell, M.J.W. Evolution of Chip-Scale Heterodyne Optical Phase-Locked Loops Toward Watt Level Power Consumption. *IEEE/OSA J. Lightwave Technol.* **2018**, *36*, 258–264. [CrossRef]
9. O'Reilly, J.J.; Lane, P.M.; Heidemann, R.; Hofstetter, R. Optical generation of very narrow linewidth millimeter wave signals. *Electron. Lett.* **1992**, *28*, 2309–2310. [CrossRef]
10. Izutsu, M.; Yanase, Y.; Sueta, T. Broadband traveling wave modulator using a $LiNbO_3$ optical waveguide. *IEEE J. Quantum Electron.* **1977**, *QE-13*, 287–290. [CrossRef]
11. Izutsu, M.; Itoh, T.; Sueta, T. 10 GHz bandwidth traveling-wave $LiNbO_3$ optical waveguide modulator. *IEEE J. Sel. Top. Quantum Electron.* **1978**, *QE-14*, 394–395. [CrossRef]
12. Mohamed, M.; Zhang, X.; Hraimel, B.; Wu, K. Analysis of frequency quadrupling using a single Mach-Zehnder modulator for millimeter-wave generation and distribution over fiber systems. *Opt. Express* **2008**, *16*, 10786–10802. [CrossRef] [PubMed]
13. Kawanishi, T.; Sasaki, M.; Shimotsu, S.; Oikawa, S.; Izutsu, M. Reciprocating optical modulation for harmonic generation. *IEEE Photon. Technol. Lett.* **2001**, *13*, 854–856. [CrossRef]
14. Pan, Z.; Chandel, S.; Yu, C. 160 GHz optical pulse generation using a 40 GHz phase modulator and two stages of delayed MZ interferometers. In Proceedings of the Technical Digest of Conference on Lasers and Electro-Optics and 2006 Quantum Electronics and Laser Science Conference 2006 (CD), Long Beach, CA, USA, 21–26 May 2006; Optical Society of America: Washington, DC, USA, 2006. Paper CFP2. pp. 1–2.

15. Izutsu, M.; Shikama, S.; Sueta, T. Integrated Optical SSB Modulator/Frequency Shifter. *IEEE J. Quantum Electron.* **1981**, *QE-17*, 2225–2227. [CrossRef]
16. Shimotsu, S.; Oikawa, S.; Saitou, T.; Mitsugi, M.; Kurodera, K.; Kawanishi, T.; Izutsu, M. Single side-band modulation performance of a $LiNbO_3$ integrated modulator consisting of four-phase modulator waveguides. *IEEE Photon. Technol. Lett.* **2001**, *13*, 364–366. [CrossRef]
17. Kawanishi, T.; Sakamoto, T.; Tsuchiya, M.; Izutsu, M. High Carrier Suppression Double Sideband Modulation Using an Integrated $LiNbO_3$ Optical Modulator. In Proceedings of the 2005 International Topical Meeting on Microwave Photonics, Seoul, Korea, 14 October 2005; IEEE: Piscataway, NJ, USA, 2005; pp. 29–32.
18. Lin, C.-T.; Shih, P.-T.; Chen, J.; Jiang, W.-J.; Dai, S.-P.; Peng, P.-C.; Ho, Y.-L.; Chi, S. Optical millimeter-wave up-conversion employing frequency quadrupling without optical filtering. *IEEE Trans. Microw. Theory Tech.* **2009**, *57*, 2084–2092.
19. Lin, C.-T.; Shih, P.-T.; Jiang, W.-J.; Chen, J.J.; Peng, P.-C.; Chi, S. A continuously tunable and filterless optical millimeter-wave generation via frequency octupling. *Opt. Express* **2009**, *17*, 19749–19756. [PubMed]
20. Shi, P.; Yu, S.; Li, Z.; Song, J.; Shen, J.; Qiao, Y.; Gu, W. A novel frequency sextupling scheme for optical mm-wave generation utilizing an integrated dual-parallel Mach-Zehnder modulator. *Opt. Commun.* **2010**, *283*, 3667–3672. [CrossRef]
21. Lu, G.W.; Sakamoto, T.; Chiba, A.; Kawanishi, T.; Miyazaki, T.; Higuma, K.; Ichikawa, J. Optical minimum-shift-keying transmitter based on a monolithically integrated quad Mach–Zehnder in-phase and quadrature modulator. *Opt. Lett.* **2009**, *34*, 2144–2146. [CrossRef]
22. Kaneko, A.; Yamazaki, H.; Yamada, T. Compact Integrated 100 Gb/s Optical Modulators Using Hybrid Assembly Technique with Silica-Based PLCs and $LiNbO_3$ Devices. In Proceedings of the Optical Fiber Communication Conference and National Fiber Optic Engineers Conference, OSA Technical Digest (CD), San Diego, CA, USA, 22–26 March 2009; Optical Society of America: Washington, DC, USA, 2009. Paper OThN3. pp. 1–2.
23. Chiba, A. Integrated optical modulation devices open the door for optical communication being close to the Shannon limit. *Electron. Lett.* **2010**, *46*, 186. [CrossRef]
24. Chiba, A.; Sakamoto, T.; Kawanishi, T.; Higuma, K.; Sudo, M.; Ichikawa, J. 16-level quadrature amplitude modulation by monolithic quad-parallel Mach-Zehnder optical modulator. *Electron. Lett.* **2010**, *46*, 227–228. [CrossRef]
25. Sakamoto, T.; Chiba, A. Coherent synthesis of optical multilevel signals by electrooptic digital-to-analog conversion using multiparallel modulator. *J. Sel. Top. Quantum Electron.* **2010**, *16*, 1140–1149. [CrossRef]
26. Liu, W.; Wang, M.; Yao, J. Tunable Microwave and Sub-Terahertz Generation Based on Frequency Quadrupling Using a Single Polarization Modulator. *IEEE/OSA J. Lightwave Technol.* **2013**, *31*, 1636–1644.
27. Wang, W.T.; Li, W.; Zhu, N.H. Frequency quadrupling optoelectronic oscillator using a single polarization modulator in a Sagnac loop. *Opt. Commun.* **2014**, *318*, 162–165. [CrossRef]
28. Chiba, A.; Akamatsu, Y.; Takada, K. Optical two-tone signal generation without use of optical filter for photonics-assisted radio-frequency quadrupling. *Opt. Lett.* **2015**, *40*, 3651–3654. [CrossRef] [PubMed]
29. Chiba, A.; Akamatsu, Y.; Takada, K. RF frequency sextupling via an optical two-tone signal generated from two modulation lightwaves from one Mach-Zehnder optical modulator. *Opt. Express* **2015**, *23*, 26259–26267. [CrossRef] [PubMed]
30. Chiba, A.; Akamatsu, Y.; Takada, K. Long-term stable 60 GHz optical two-tone signal by destructive optical interference obtained from RF phase adjustment. *Electron. Lett.* **2016**, *52*, 736–737. [CrossRef]
31. Chiba, A.; Akamatsu, Y.; Takada, K. Wavelength- independent optical two-tone signal generator composed of one single Mach-Zehnder optical modulator and a polarizer. In Proceedings of the 2016 IEEE Photonics Conference, Waikoloa, HI, USA, 2–6 October 2016; IEEE: Piscataway, NJ, USA, 2009; pp. 736–738.

Article

Broadband Generation of Polarization-Immune Cloaking via a Hybrid Phase-Change Metasurface

Ximin Tian [1], Junwei Xu [1], Ting-Hui Xiao [2], Pei Ding [1], Kun Xu [1], Yinxiao Du [1,]* and Zhi-Yuan Li [3]

1 School of Materials Science and Engineering, Zhengzhou University of Aeronautics, Zhengzhou 450046, China; xmtian007@zua.edu.cn (X.T.); xujunwei001@zua.edu.cn (J.X.); dingpei@zua.edu.cn (P.D.); kunxu@zua.edu.cn (K.X.)
2 Department of Chemistry, University of Tokyo, Tokyo 113-0033, Japan; xiaoth@chem.s.u-tokyo.ac.jp
3 College of Physics and Optoelectronics, South China University of Technology, Guangzhou 510640, China; phzyli@scut.edu.cn
* Correspondence: duyinxiao@zua.edu.cn

Abstract: Metasurface-enabled cloaking offers an alternative platform to render scatterers of arbitrary shapes indiscernible. However, specific propagation phases generated by the constituent elements for cloaking are usually valid for a single or few states of polarization (SOP), imposing serious restrictions on their applications in broadband and spin-states manipulation. Moreover, the functionality of a conventional metasurface cloak is locked once fabricated due to the absence of active elements. Here, we propose a hybrid phase-change metasurface carpet cloak consisting of coupled phase-shift elements setting on novel phase-change material of $Ge_2Sb_2Se_4Te_1$ (GSST). By elaborately arranging meta-atoms at either 0 or 90 degrees on the external surface of the hidden targets, the wavefront of its scattered lights can be thoroughly rebuilt for arbitrary SOP exactly as if the incidence is reflected by a flat ground, ensuring the targets' escape from polarization-scanning detections. Furthermore, the robustness of phase dispersion of meta-atoms endows the metasurface cloak wideband indiscernibility ranging from 7.55 to 8.35 μm and tolerated incident angles at least within ±25°. By reversibly switching of the phase states of $Ge_2Sb_2Se_4Te_1$, the stealth function of our design can be turned on and off. The generality of our approach will provide a straightforward platform for polarization-immune cloaking, and may find potential applications in various fields such as electromagnetic camouflage and illusion and so forth.

Keywords: metasurface carpet cloak; states of polarization (SOP); hybrid phase-change metasurface; polarization-insensitive cloaking; electromagnetic camouflage and illusion

Citation: Tian, X.; Xu, J.; Xiao, T.-H.; Ding, P.; Xu, K.; Du, Y.; Li, Z.-Y. Broadband Generation of Polarization-Immune Cloaking via a Hybrid Phase-Change Metasurface. *Photonics* **2022**, *9*, 156. https://doi.org/10.3390/photonics9030156

Received: 17 January 2022
Accepted: 1 March 2022
Published: 4 March 2022

1. Introduction

The electromagnetic (EM) invisibility, by eliminating the scattering light and restoring the polarization, amplitude and phase profiles of the transmitted/reflected light as if the incidence is reflected by a flat ground, can render the hidden targets unobservable, which is a dream that people have been chasing unremittingly for a long time. Such scenarios, often considered to only exist in magical movies by exploiting a specific invisible technique, such as in the Harry Potter movies, have become a reality and have flourished because of the advent of metamaterials [1,2]. Transformation optics (TO) [3], as an ambitious approach, can reboot the EM waves around the hidden objects and thereby enable it to be fully indiscernible. However, the implementation of this technique is supported by complicated constitutive parameters, especially for the extremely inhomogeneous and anisotropic profiles of the material parameters [4–6], which impose enormous challenges in practice. Afterward, the concept of carpet cloak based on scattering cancelation was proposed [7,8], which is fulfilled by suppressing the dominant multipolar scattering orders. Nevertheless, bulky footprint and lateral shift of the scattered waves can easily deteriorate the quality of cloaking and thus render the hidden objects readily detectable, especially at high frequencies. As an alternative, the cloak technique assisted by quasi-conformal mapping was also proposed [9,10], which adopts the strategy of inverse transformation of the permittivity and permeability to release the inherent restraint of full stealth technique. However, it is generally bulky, costly and time consuming to manufacture with high precision.

Recently, the emergence of metasurfaces [11–16] undoubtedly provides a powerful platform to achieve stealth, due to its inherent virtues such as more degrees of flexible-design freedom, skin thickness, low loss and easy fabrication relative to three-dimensional (3D) bulky metamaterials. Metasurfaces, as the equivalent 2D counterpart of metamaterials, consist of subwavelength optical scatterers that could arbitrarily tailor the polarization, phase and amplitude of EM waves by introducing abrupt phase changes across the interface. Metasurface-enabled carpet cloak [17–23] inherits all the merits of metasurfaces and any targets wrapped with it can thoroughly reconstruct the scattered lights exactly as if the input EM wave is reflected by a flat conducting ground. Following this strategy, various types including multi-wavelength [24,25], reconfigurable [26,27] and smart metasurface carpet cloaks [28–30] have been proposed and demonstrated. However, these metasurface cloaks are subjected to polarization sensitivity, i.e., they work efficiently only at a single or a few states of polarization (SOP), being easily detectable by full-polarization detection systems. Such restrictions can be released by exploiting symmetric circular ring or square-shaped nanostructures as the constituent elements of metasurfaces [19,31], but at the cost of losing a degree of freedom in the design space. On the other hand, the functionality of current metasurface cloaks is locked once fabricated due to the absence of active elements, which severely limits their wide applications. As a non-volatile optical phase change material (O-PCM), $Ge_2Sb_2Se_4Te_1$, with extremely large refractive index contrast associated with material phase transformation as well as exceptionally broadband transparency and low loss in the infrared spectral regime, uniquely empowers metasurface devices with more degrees of freedom for post-processing and thus to become active cloaks. Thereafter, a reconfigurable full-polarization metasurface cloak has been in high demand but challenging to develop until now.

Here, counterintuitively, we report an ambitious approach to construct a metasurface consisting of anisotropic Au nanoantenna pairs setting on a $Ge_2Sb_2Se_4Te_1$ spacer layer, which is capable of perfect cloaking at any SOP described by the Poincaré sphere. In contrast to [21], who realized polarization-insensitive cloaking via exploiting the cross-LP scheme, in our scheme these anisotropic nanoantenna pairs with varied dimensions are arranged at either 0 or 90 degrees (relative to the x-axis), allowing us to accurately implement the propagation phases to cover 2π phase shifts. There are two advantages by doing so. First, both right circularly polarized (RCP) and left circularly polarized (LCP) light will be imparted by the identical phase profiles as they interact with the designed metasurface, leading to a perfect cloaking performance upon any incident SOP since any polarized light can be decomposed into a combination of LCP and RCP light. Second, the robustness of phase dispersions of these anisotropic meta-atoms endows the metasurface cloak wideband indiscernibility and large tolerated angular range. One point should be emphasized that by reversibly switching of phase states of $Ge_2Sb_2Se_4Te_1$, the designed cloak can realize stealth switching of "ON" and "OFF" by imposing appropriate external stimuli without changing the meta-devices' structures. The generality of our approach will provide a straightforward platform for reconfigurable polarization-immune cloaking, and may find potential applications in various fields such as electromagnetic camouflage and illusion and so forth.

2. Principles and Structures

2.1. Design Principles of Polarization-Insensitive Cloaking

To construct a metasurface carpet cloak that works well for metallic bumps with arbitrary boundaries of $h(x, y)$, each constituent element should be encoded with the desired compensated phase profiles $\varphi(x, y)$ aimed at the wavelength λ_0 for a specific polarization [18,21]:

$$\varphi(x,y) = \pi - 2k_0 cos\theta(h(x,y) - g(x,y)) \quad (1)$$

where $k_0 = 2\pi/\lambda_0$ is the wave number in free space, θ is the angle of input beams, $g(x, y)$ denotes the contour of a flat ground and the additional phase π is induced by half wave loss. Generally, one can use the function $g(x, y)$ to mimic any fictive object as an illusion cloak. For simplicity, $g(x, y)$ is set to 0 to signify a flat ground herein. Note that the compensated phase profiles $\varphi(x, y)$ are only feasible for a specific polarization to realize perfect stealth. Other SOP deteriorate the entire cloaking performance due to unwanted phase distortions, which can be overcome by employing a symmetric circular ring or square-shaped nanostructures as the meta-atoms of metasurfaces. However, this approach suffers from the loss of a degree of freedom in the design space owing to the symmetry of these meta-atoms.

To solve this issue, we counterintuitively adopt anisotropic Au nanoantenna pairs as the meta-atoms of metasurfaces, which synergies propagation phases tailored by the cells' dimensions and specific PB phases by rotating meta-atoms at either 0 or 90 degrees, similar to those of our pioneered work [32] and Chen's work [33]. To be specific, as the input spin-polarized light hits on anisotropic meta-atoms, the reflected beams could be described by the Jones vector [34]:

$$R|\sigma\rangle = \frac{r_l + r_s}{2}|\sigma\rangle + \frac{r_l - r_s}{2}exp(-j2\sigma\beta)|-\sigma\rangle \quad (2)$$

in which the parameter σ is assigned to +1 or -1 for RCP or LCP light, and r_l and r_s represent the complex reflection coefficients for the linearly polarized (LP) light along the long and short axes of the nanoantenna, respectively. Each nanoantenna is rotated counter-clockwise with the angle of β with respect to the x-axis. From Equation (2), we can find that the outgoing EM fields contain not only the original (co-) polarized components (σ) with a complex amplitude $(r_l + r_s)/2$, but also the orthogonal (cross-) polarized components ($-\sigma$) with a complex amplitude $(r_l - r_s)/2$. It should be noted that an additional phase delay $2\sigma\beta$ is induced for the cross-polarized components, which is the origin of the PB phase. We can eliminate the unfavorable scatterings caused by the co-polarized components via optimizing the unit cells to be a perfect half-wave plates ($r_l = -r_s$), while maximizing the cross-polarized components to boost the polarization conversion efficiency (PCE) and realize high-efficiency meta-devices. However, since the PB phase is susceptible to the helicity of the incident beams, i.e., $\exp(j2\beta)$ for RCP and $\exp(-j2\beta)$ for LCP lights, respectively, the above cross-polarization-engineered phases are only feasible for a specific polarization, while collapsing for others. Notably, if half of all the nanoantennas are arranged at $\beta = 0°$ and the other half at $\beta = 90°$, the exponential expression results will be identical, manifesting the same phase distributions for input RCP or LCP beams, which implies that the metasurface will enable perfect cloaking upon any incident SOP.

2.2. Structures and Methods

Figure 1a illustrates the schematic of the polarization-immune metasurface carpet cloak, from which one could witness that the metasurface design enables perfect cloaking upon all LP (transverse magnetic, TM), LP (transverse electric, TE), LP($\pi/4$), LCP and RCP incident beams. These prototypical space-variant polarization states of incident beams can be gracefully described by the Poincaré sphere (PS) [35], as shown in Figure 2a. As a proof, an ultrathin metasurface cloak is designed, which overlays a triangular bump, as exhibited in Figure 1b. The tilt angle of the triangular bump is $a = 15°$, and $h_i = (i-1/2)\,p$ represents the vertical distance from the center point of the meta-atom (i = 1,2,3, ... , n) to the flat ground. For better cloaking, 40 meta-atoms (n = 40) are arranged on each side of the triangular bump. Since the tilt angles of the two sides of the bump exhibit opposite signs, the deflection angles of reflected beams are also inverted, leading to mirror-like reflection. Left panel of Figure 1c exhibits the schematic of unit cell of the metasurface cloak, in which the spacer layers of GSST and ZnS:SiO_2 are sandwiched by top metallic nanoantenna pairs and bottom gold ground. The GSST layer with the fixed thickness of 490 nm functions basically as active elements to tailor local circumstances, and the ZnS:SiO_2 layer with a fixed thickness of 150 nm acts as the protective layer to prevent GSST from being oxidized, respectively. The top metallic nanoantenna pairs consist of two distinct anisotropic Au nanoantennas with identical height h = 100 nm and lattice constant p = 3000 nm. It should be emphasized that the 3D unit cell is arranged by 0 degrees herein. To break structural symmetry and generate chirality-dependent optical responses, one antenna A0 with a fixed width a_0 = 550 nm and length b_0 = 1200 nm orients obliquely with a rotation angle γ = 20° relative to the x-axis. While for the other A1, its long axis is parallel to the y-axis, and its width is set to a_1 = 800 nm and length b_1 varies ranging from 600 to 2900 nm to realize 0~π coverage. Additionally, Figure 1c (right panel) also displays the top views of the meta-atoms arranged by 0 (top) and 90 (bottom) degrees.

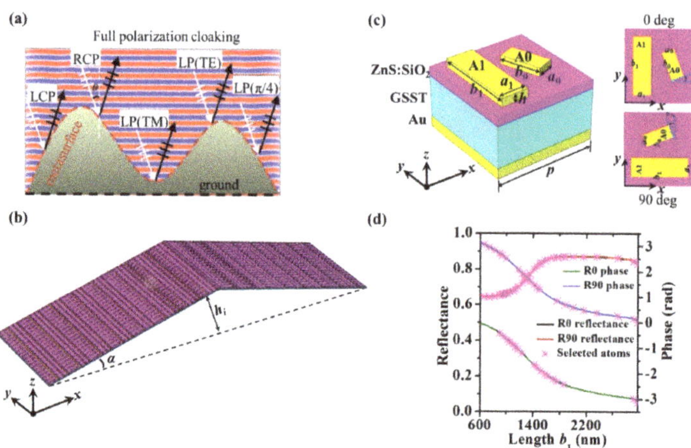

Figure 1. Principle diagram and design scheme of the polarization-immune metasurface carpet cloak. (**a**) Principle diagram of the polarization-immune metasurface carpet cloak wrapping on metallic bumps with arbitrary boundaries. It can work well upon all LP (TM), LP (TE), LP (π/4), LCP and RCP incident beams. (**b**) As a proof, an ultrathin triangular metasurface carpet cloak is designed in which the tilt angle of the triangular bump is a = 15°, h_i represents the vertical distance from the center point of the meta-atom to the flat plane. (**c**) Left: Schematic of the meta-atoms (0 degrees) of the proposed metasurface. Right: Top views of the meta-atoms arranged at 0 (top) and 90 (bottom) degrees. (**d**) Normalized reflectance and propagation phases of anisotropic Au nanoantenna pairs with length b_1 spanning from 600 to 2900 nm, while other parameters remain unchanged.

In the proof-of-concept demonstrations, finite-element-method-based numerical simulations are performed by utilizing the software of COMSOL Multiphysics. For each meta-atom, periodic boundary conditions (PBCs) are adopted along the x- and y-axes, and perfectly matched layer (PML) boundary conditions are implemented in the z direction. The input RCP (LCP) wave propagates along the $-z$-axis and the reflection spectra and propagation phase with the flipped spin states (LCP (RCP)) are captured. While for the metasurface cloak, we set PML around the model in x–z plane, apply PBCs along y-axis and adopt perfect electrical conductor (PEC) boundary instead of bottom Au ground, respectively. The incident EM waves are characterized by the incident background scattered fields that propagate along the $-z$-axis. The optical constants of GSST exhibit frequency-dependent characteristic [36], the refractive index of ZnS:SiO_2 is set to 2 [37] and the permittivity of Au can be represented by the Drude model [38]. A prototype of the metasurface cloak can be fulfilled by a standard lithography process, in which 490 nm GSST and 150 nm ZnS:SiO_2 films are DC-magnetron sputter-deposited in sequence followed by electron beam lithography, and finally 100 nm Au film is thermally evaporated.

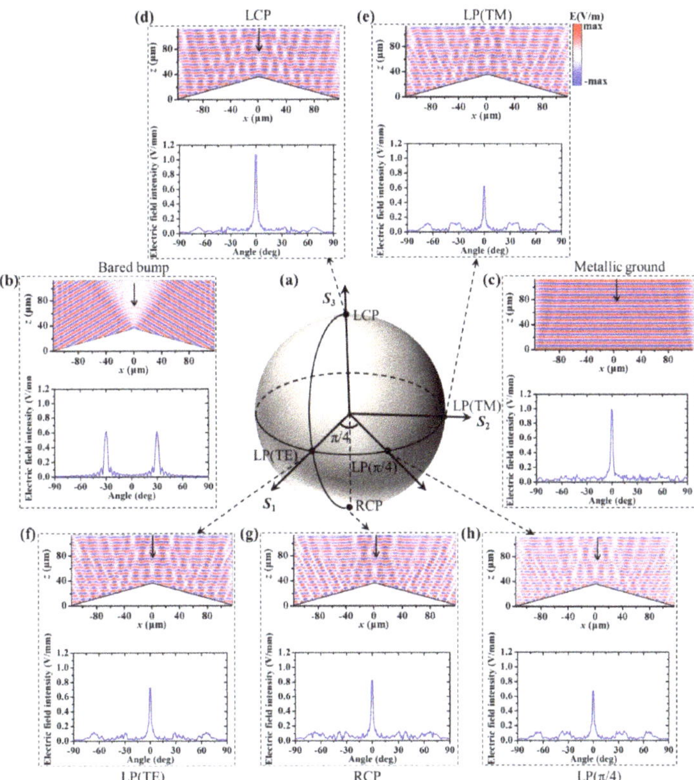

Figure 2. A Poincaré sphere representation of cloaking response under arbitrary SOP for normal incidence. (**a**) A Poincaré sphere and five points that characterize different SOP. The calculated reflected spatial E-field patterns in the x–z plane (up panels) and the corresponding farfield radiation patterns (bottom panels) for the bared bump under LCP incidence (**b**), metallic ground under LCP incidence (**c**) and the cloaked bump under LCP (**d**), LP (TM) (**e**), LP (TE) (**f**), RCP (**g**) and (**h**) LP ($\pi/4$) input light, respectively.

Figure 1d shows the simulated propagation phase shifts and reflection at the targeted wavelength of 7.9 μm by parameter sweeping of the long axis' length b_1 of A1 nanoantenna by varying it from 600 to 2900 nm for two perpendicular-oriented meta-atoms R0 and R90. In our simulation, R0, i.e., the meta-atoms arranged by 0 degrees, represents the long axis of A1 (b_1) oriented along the y-axis, while R90, i.e., the meta-atoms arranged by 90 degrees, represents the long axis of A1 (b_1) oriented along the x-axis by rotating A1 in-plane 90 degrees. It should be emphasized that the whole element structure maintains exactly the same rotation as the long axis of A1 (b_1) with reference to its center in the x–y plane. As illustrated in Figure 1d, for the meta-atoms arranged at either R0 or R90, the reflection spectra are completely coincident, while the phase shifts preserve the same profiles but are imparted with a π phase shift, which confirms the feasibility of the theory expressed by Equation (2). As a proof, 40 units that cover 2π full-phase shifts while maintaining high reflectance are picked to build the metasurface cloak, in which 22 (close to half of 40) meta-atoms are arranged in R0 and 18 in R90, enabling the realization of polarization-insensitive cloaking performance.

3. Results and Discussions

With the above principle, we designed a prototype of an ultrathin metasurface carpet cloak utilizing anisotropic nanantenna pairs as building blocks and wrapping it on a triangular bump, as shown in Figure 1b. Firstly, to characterize its polarization-immune cloaking performance at normal incidence, five points that represent different SOP of incident light impinging on the designed metasurface are picked and their coordinates at the Poincaré sphere are given in Figure 2a. For the Poincaré sphere, the North (up) and South (bottom) poles denote the cases of two orthogonal circularly polarized illuminances, and Points B, C and D on the equator represent the scenarios upon LP lights with polarization angles 0 (TE), $\pi/4$ and $\pi/2$ (TM), respectively. The corresponding calculated reflected spatial E-field patterns in the x–z plane are illustrated in the up panels of Figure 2d–h, respectively. For comparison, the cases of a bared triangular bump and PEC ground upon LCP illuminance are also considered, as shown in the up panels of Figure 2b,c, respectively. Compared with the PEC ground, tangible distortions of the scattering field are induced by the bared bump, and the wavefront is substantially distributed in parallel with double tilt angle α (Figure 2b), very different from the case of PEC ground, for which the wavefront is parallel to the ground plane (Figure 2c). However, after wrapping the bared bump with our cleverly designed metasurface, when LCP light illuminates, scattering distortions are strongly suppressed and the wavefront is reconstructed (Figure 2d) as those of the ground plane (Figure 2c). As a consequence, the hidden target is invisible as if it does not exist. It is worth mentioning that almost identical restored wavefronts are also yielded for our cleverly designed metasurface upon all the other illuminances of RCP, LP (TM), LP (TE) and LP ($\pi/4$), implying that the metasurface cloak enables perfect cloaking upon any incident SOP. The existing little perturbation is probably due to the discontinuity of local reflection phases, which can be minimized by further optimization.

To better quantitively evaluate the quality of the generated cloaking performance by our metasurface cloak, we show the corresponding far-field radiation patterns upon the above-mentioned incident SOP, as shown in the bottom panels of Figure 2d–h. Simultaneously, the corresponding far-field radiation patterns for the bared bump and PEC ground are also depicted in the bottom panels of Figure 2b,c for comparison. As expected, two scattered sidelobes located at $\pm 30^\circ$ (double of the tilt angle α) produced by the bared bump are efficiently suppressed and merged into one main lobe by our metasurface cloak upon any incident SOP. Such a phenomenon is very consistent with those of PEC ground, again proving that our designed metasurface enables polarization-insensitive cloaking performance at the wavelength of 7.9 μm upon normal incidence.

Furthermore, oblique incidence scenarios at 25° for the metasurface cloak upon the above-mentioned five SOP are calculated, as shown in Figure 3c–g. We also give the cases of bared cloak and PEC ground for comparison. It can be easily found that the oblique incidence of 25° breaks the symmetry of the scattered wavefront between the left and right sides for the bared bump (Figure 3a), quite different from the specular reflection of the PEC ground (Figure 3b). To our relief, after the bared bump was wrapped with the well-designed metasurface, the reflection behaviors at 25° upon all five SOP (Figure 3c–g) were analogous to those of the PEC ground, explicitly implying that the designed metasurface cloak is insensitive to the incident SOP even at modest oblique angles, and straightforwardly proving that the unique design could tolerate incident oblique angles of at least 25 degrees at the wavelength of 7.9 μm.

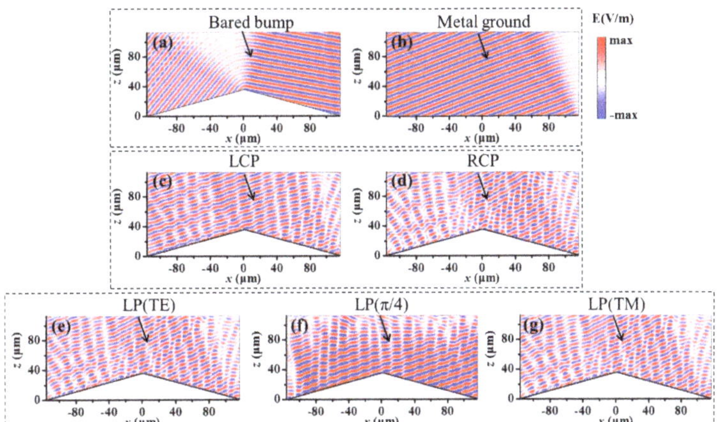

Figure 3. Cloaking response of our design upon arbitrary SOP under oblique incidence. For our design under the oblique incidence of 25°, the simulated reflected spatial distributions in the *x*–*z* plane for the bared bump under LCP incidence (**a**), metallic ground under LCP incidence (**b**) and the cloaked bump under LCP (**c**), RCP (**d**), LP (TE) (**e**), LP ($\pi/4$) (**f**) and LP (TM) (**g**) illuminance, respectively.

Broadband cloaking operation is a crucial indicator to evaluate the performance of invisibility cloaks. For convenience, we only showed the simulated far-field radiation profiles of the cloaked bump upon normal LCP incidence against the input wavelength from 7 to 9 μm in Figure 4a. One could see that the far-field radiation profiles are converged into a straight horizontal line located at 0 degrees within a wideband from 7.55 (λ_1) to 8.35 μm (λ_2) (labelled by dashed white lines), i.e., our metasurface cloak exhibits a broadband cloaking response despite some trivial scattering perturbations. To reveal the underlying mechanism, we calculated the theoretical phase dispersions of all the picked meta-atoms according to Equation (1) under three prototypical incident wavelengths (7.55, 7.9 and 8.35 μm) in Figure 4b. The actual (simulated) phase dispersions for all selected meta-atoms at 7.9 μm are also displayed. One could observe that the theoretical predictions exactly match with the actual phases at λ_0 = 7.9 μm, leading to invulnerable invisibility. Although the theoretical phase lines at λ_1 = 7.55 μm and λ_2 = 8.35 μm slightly deviate from the actual ones, the designed metasurface cloak can still preserve perfect invisibility due to the robustness of the phase dispersion of meta-atoms [39]. Figure 4d,e shows the simulated reflected E-field intensity profiles in the *x*–*z* plane and their corresponding 2D far-field radiation patterns for our scheme under LCP illuminance with λ_1 = 7.55 μm and λ_2 = 8.35 μm, respectively. As expected, these reconstructed wavefronts exhibit almost identical morphological features to the one at the wavelength of 7.9 μm, unanimously confirming that the metasurface can obtain broadband and invulnerable invisibility in the MIR. To further verify the broadband cloaking operation of our design, Figure 4c depicts the reduced total radar cross-section (RCS) that is calculated by dividing the total RCS of the cloaked bump by that of the bared bump [40]. It can be easily found that our design upon normal incidence produced a significant RCS reduction reaching up to −18 dB, and the 3 dB RCS reduction bandwidth is very consistent with the results in Figure 4a, further implying the feasibility of the broadband cloaking operation of the designed metasurface. Additionally, the relatively large RCS reduction upon oblique incidence (θ = 25°) demonstrates the virtues of wide-angle cloaking very well.

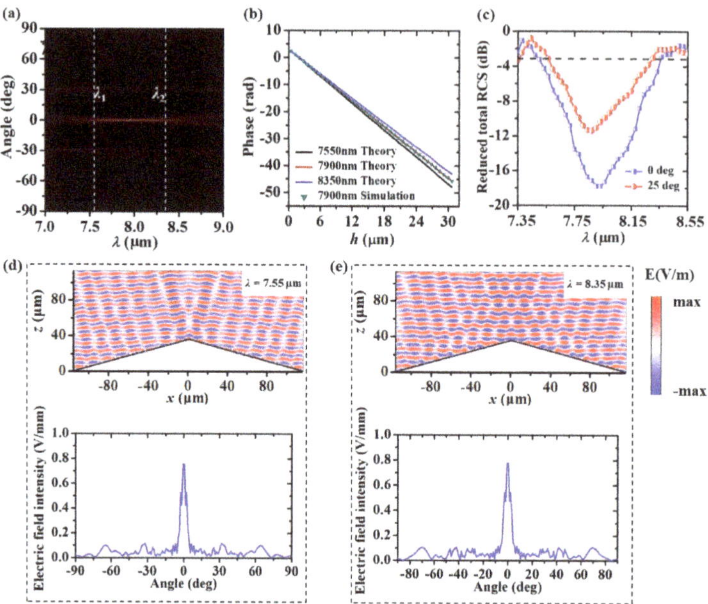

Figure 4. Broadband cloaking operation of the polarization-insensitive metasurface carpet cloak. (**a**) Scanned plot of the simulated far-field radiation profiles for the cloaked bump upon LCP incidence. (**b**) Phase dispersions of all picked meta-atoms under three prototypical incidences. (**c**) Calculated RCS reduction for our metasurface cloak under normal and oblique incidences. Simulated reflected E-field intensity profiles in the $x-z$ plane (up panels) and the corresponding 2D far-field radiation patterns (bottom panels) for our scheme under LCP illuminance with λ_1 = 7.55 μm (**d**) and λ_2 = 8.35 μm (**e**), respectively.

It should be emphasized that our metasurface cloak is endowed with reconfigurability and continuous (multilevel) control supported by embedding a chalcogenide phase-change layer of GSST as a spacer layer. As a representative chalcogenide PCM, GSST could be switched instantaneously, repeatedly and non-volatilely along amorphous (aGSST) intermediate and crystalline (cGSST) states [41]. A huge contrast in the complex refractive index caused by phase transformation and low loss in the MIR enable GSST to be a good platform for reconfigurable and versatile optics [36]. To confirm this fact, Figure 5 displays the simulated reflected E-field intensity profiles in the x–z plane for our scheme with different crystallization levels (m = 0.25, 0.5, 0.75 and 1) upon LCP illuminance of λ_0 = 7.9 μm, respectively. For the calculation of the effective permittivity of GSST at any crystallization level, please refer to Ref. [42]. m = 0 and m = 1 indicate that GSST exists in amorphous and crystalline states, respectively, and the larger the value of m, the higher the crystallinity of GSST. This work adopts aGSST by default, unless otherwise specified. The results of the scenario with m = 0 have been shown in Figure 2d. As anticipated, the stealth performance of the designed metasurface cloak gradually deteriorates as m increases, which is mainly attributed to the red shift of resonance and enlarged loss. When m = 1 (Figure 5d), the reflected wavefront of our design looks very similar to that of the bared bump (Figure 2b), indicating that the designed metasurface cloak turns off. Therefore, by adjusting the crystallinity of GSST, our designed metasurface cloak can realize the function of continuous regulation and switching, which undoubtedly expands the flexibility and diversity of the designed functionalities.

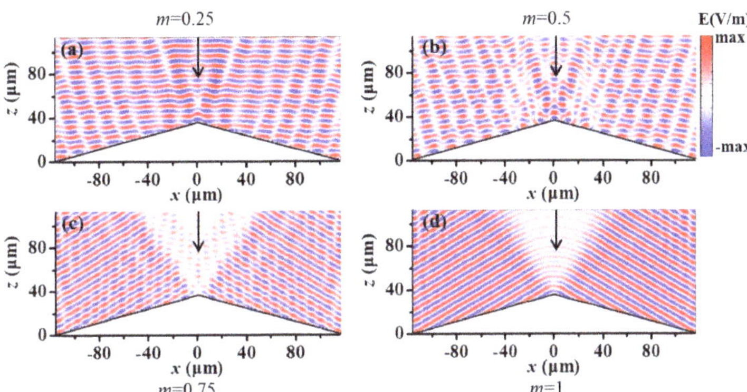

Figure 5. GSST state-dependent cloaking performance of our design under normal incidence. Simulated reflected E-field distributions in the x–z plane for our scheme with different crystallization levels of m = 0.25 (**a**), 0.5 (**b**), 0.75 (**c**) and 1 (**d**) upon LCP illuminance with λ_0 = 7.9 μm nm, respectively.

4. Conclusions

In summary, we report a $Ge_2Sb_2Se_4Te_1$-based metasurface carpet cloak consisting of 40 anisotropic nanostructures. By arranging well-optimized meta-atoms at either 0 or 90 degrees on the external surface of the hidden targets (triangular bared bump), 2π full-phase coverage can be obtained, ensuring the wavefront of scattered lights can be thoroughly rebuilt exactly as if the incidence is reflected by a flat ground under arbitrary SOP of the Poincaré sphere, and thus enabling the hidden targets to escape from polarization-scanning detections. Meanwhile, the robustness of the phase dispersion of meta-atoms allows for the metasurface cloak to generate a broadband indiscernibility within the wavelength range of 7.55 to 8.35 μm and to tolerate the incident angles reaching up to $\pm 25°$. It should be emphasized that our metasurface cloak is endowed with a continuously adjustable and switchable stealth response supported by converting GSST states. The generality of our approach will provide a straightforward platform for polarization-immune cloaking, and may find potential applications in various fields such as electromagnetic camouflage and illusion and so forth.

Author Contributions: Conceptualization, X.T. and J.X.; methodology, X.T.; software, X.T. and J.X.; formal analysis, X.T. and J.X.; investigation, X.T., J.X. and T.-H.X.; data curation, X.T. and J.X.; writing—original draft preparation, X.T. and J.X.; writing—review and editing, X.T., J.X., K.X. and Y.D.; supervision, X.T., Y.D. and Z.-Y.L.; project administration, X.T., J.X., Y.D. and P.D. All authors have read and agreed to the published version of the manuscript.

Funding: This research was funded by the National Natural Science Foundation of China (No. 12004347); the Scientific and Technological Project in Henan Province (Nos. 212102310255, 202102310535, 222102210063); the Aeronautical Science Foundation of China (Nos. 2020Z073055002, 2019ZF055002); and the Innovative Research Team (in Science and Technology) in the University of Henan Province (No. 22IRTSTHN004).

Institutional Review Board Statement: Not applicable.

Informed Consent Statement: Not applicable.

Data Availability Statement: The data presented in this study are available upon request from the corresponding author. The data are not publicly available because the data also forms part of an ongoing study.

Acknowledgments: The authors express their appreciation to the anonymous reviewers for their valuable suggestions.

Conflicts of Interest: The authors declare no conflict of interest.

Article

Control of Surface Plasmon Resonance in Silver Nanocubes by CEP-Locked Laser Pulse

Ju Liu [1] and Zhiyuan Li [2,*]

1 School of Science, China University of Mining and Technology-Beijing, Beijing 100083, China; juliu67@sina.com
2 College of Physics and Optoelectronics, South China University of Technology, Guangzhou 510641, China
* Correspondence: phzyli@scut.edu.cn

Abstract: Localized surface plasmon resonance (LSPR) of metal nanoparticles has attracted increasing attention in surface-enhanced Raman scattering, chemical and biological sensing applications. In this article, we calculate the optical extinction spectra of a silver nanocube driven by an ultrashort carrier envelope phase (CEP)-locked laser pulse. Five LSPR modes are clearly excited in the optical spectra. We analyze the physical origin of each mode from the charge distribution on different parts of the cubic particle and the dipole and quadrupole excitation features at the LSPR peaks. The charge distribution follows a simple rule that when the charge concentrates from the face to the corners of the cubic particle, the resonant wavelength red-shifts. Then we modulate the LSPR spectra by changing CEP. The results show that CEP has selective plasmon mode excitation functionality and can act as a novel modulation role on LSPR modes. Our work suggests a novel means to regulate LSPR modes and the corresponding optical properties of metal nanoparticles via various freedoms of controlled optical field, which can be useful for optimized applications in chemical and biological sensors, single molecule detection, and so on.

Keywords: surface plasmons; ultrashort laser pulse; carrier envelope phase; nanoparticles; nanocubes; modulation

Citation: Liu, J.; Li, Z. Control of Surface Plasmon Resonance in Silver Nanocubes by CEP-Locked Laser Pulse. *Photonics* **2022**, *9*, 53. https://doi.org/10.3390/photonics9020053

Received: 20 December 2021
Accepted: 14 January 2022
Published: 19 January 2022

Publisher's Note: MDPI stays neutral with regard to jurisdictional claims in published maps and institutional affiliations.

1. Introduction

The optical properties of metal nanoparticles have long been of great interest in physics, chemistry, biology and their interdisciplinary fields. Localized surface plasmon resonance (LSPR), which is a collective oscillation of conduction electrons, occurs when metal nanoparticles interact with incident light waves [1]. This resonance leads to large enhancements of local electromagnetic field around the nanoparticle surface [2] and is sensitive to nanoparticle size, shape, composition and deposited substrate as well as the external dielectric environment [3]. These unique properties make the research of LSPR spectroscopy significant for application to surface-enhanced Raman scattering (SERS) [4], chemical and biological sensing [5], antennae [6,7] and so on.

For many metal nanoparticles, the optical extinction spectra present a series of LSPR peaks in a broad spectral range with different amplitudes and widths. Understanding and predicting the fundamental physics governing LSPRs is both necessary to realize and fully optimize potential devices. Mie theory can precisely calculate the LSPR modes of a spherical Ag nanoparticle with size larger than 80 nm. Higher plasmon modes appear in the optical spectrum and different modes would localize at different spatial regions of the nanoparticle [8–10]. In comparison, the research on Ag nanocubes is much less complete, as an analytical solution is not available for precise calculation of their optical spectra. Fuchs was the first to envision theoretically the response of a cubic particle described by a model dielectric constant in the electrostatic limit and predicted several surface phonon modes of ionic nanocubes [11,12]. When a cubic particle interacts with continuous wave (CW) light, one or two LSPR peaks are observed for smaller particles with sizes below

80 nm, and a third peak appears between the two peaks when the nanocube edge length is longer than 90 nm [1]. However, these LSPR peaks, including both the width and amplitude, are difficult to distinguish clearly and precisely. As a result, previous studies have mainly concentrated on analysis of the main dipole and quadrupole modes while ignoring higher-order modes [13,14].

To solve the problem and clearly identify all plasmon modes in metal nanoparticles, we consider the interaction of a nanocubic Ag nanoparticle with an ultrashort laser pulse. The emergence of ultrashort pulse laser technology [15,16] has provided a powerful tool for probing physical problems in an unprecedented fast timescale. Recent progress in ultrashort pulse laser technology has resulted in the generation of intense optical pulses comprising only a few wave cycles within the full width at half maximum (FWHM) of their temporal intensity profile [17–21]. The study of the interaction of intense few-cycle laser pulses with matter has brought a new, important branch of investigations in nonlinear optics [22] and has opened up a number of applications ranging from nanometer-scale materials processing [23] to the generation of coherent soft-X-ray radiation for biological microscopy [24]. With the FWHM of laser pulse becoming comparable to the time period of oscillation cycles, the temporal evolution of the electric and magnetic fields of a few-cycle light pulse and, hence, all nonlinear processes driven by these fields become increasingly affected by the carrier-envelope phase (CEP) of the pulse [25–28]. For instance, in 2008 Goulielmakis and coworkers measured the sub-femtosecond XUV emission from neon atoms ionized by a linearly polarized, sub-1.5-cycle, 720-nm laser field [29]. The ratio of the energy of the main attosecond XUV pulse to the overall XUV emission energy transmitting through the bandpass strongly depends on the CEP. At present, the regulation and control of LSPR is made mainly by selecting the particle size and shape, as well as the surrounding dielectric medium [30,31]. The emergence of ultrashort pulses hopefully may provide another regulative parameter, due to the CEP effect.

In this paper, we calculate the optical extinction spectra of a silver nanocubic particle driven by an ultrashort CEP-locked pulse. We analyze the physical origin of the five LSPR modes appearing in the spectra by examining the distribution patterns of electric fields and electric charges on the face of the cube. The result shows that when the charge concentrates from the face to the corners of the cubic particle, the resonant wavelength redshifts. We then investigate the influence of the CEP of laser pulse on the regulation and control of the LSPR modes.

2. Materials and Methods

The optical extinction spectra of a silver nanocubic particle is calculated by three-dimensional (3D) finite-difference time-domain (FDTD) simulations [32,33]. For comparison, we consider the interaction of a 3D silver nanocubic particle of edge length 90 nm with a CW light and an ultrashort CEP-locked pulse. The particle, as schematically depicted in Figure 1a, is embedded within an air background with a refractive index of 1.0. We choose the center of the silver particle as the origin of coordinates, and the edges of the nanocube are parallel to the x, y and z axes. The incident light propagates along the z axis and the polarization direction is along the x axis. We adopt the optical constants given in Palik [34] and calculate the extinction cross section of Ag nanoparticle by using the 3D FDTD method. In our calculation, a large simulation span of 3 um and a long simulation time of 800 fs are used. The simulation region uses PML absorbing boundary conditions on all boundaries and the maximum mesh step is 2 nm. To calculate the extinction spectra, we set an analysis group located inside the source to measure the absorption cross-section and an analysis group located outside the source to measure the scattering cross-section. Meanwhile, a two-dimensional monitor is set in the xz plane to calculate the electric field intensity and an analysis group with span of 100 nm to calculate the charge density distribution.

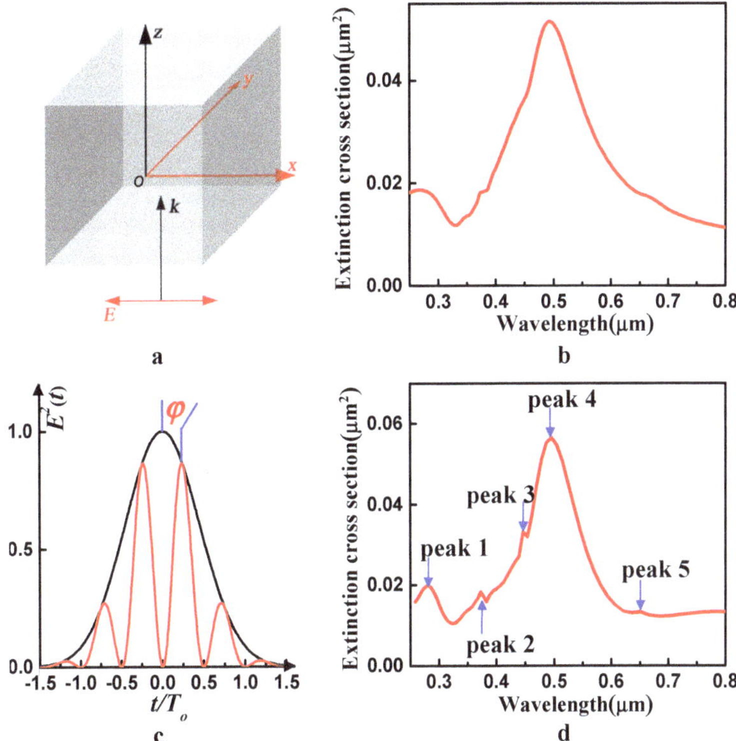

Figure 1. (**a**) Schematic configuration of the nanocube and the coordinate system for the simulation. (**b**) Calculated optical extinction cross section spectra for the Ag nanocube embedded in an air background with an edge length of 90 nm driven by a continuous wave light. (**c**) Schematic of an ultrashort pulse. φ is the CEP, T_0 is the time period of the laser field. (**d**) Calculated optical extinction cross section spectra for the Ag nanocube embedded in an air background with an edge length of 90 nm driven by an ultrashort pulse.

3. Results

We first consider the interaction of a silver nanocubic particle with a CW light. The result is shown in Figure 1b. We find that only one dominant LSPR peak located at 494 nm can be observed obviously in the spectrum, whereas other peaks are hardly recognizable and may overlap with each other. Previous studies focus on analysis of the contributions of the dipole component and quadrupole component of the dominant peak to the optical cross sections, while much less attention has been paid to the other plasmon modes when driven by a CW light.

To solve the problem and figure out the plasmon modes of the silver particle clearly, we employ a CEP-locked ultrashort pulse to interact with the silver nanocubic particle, as illustrated in Figure 1c. The electric field of the laser pulse is written as

$$E(t) = E_0 \exp(-\alpha^2 t^2) \cos(\omega t + \varphi) \tag{1}$$

where E_0 is the amplitude, $\alpha = \sqrt{2\ln(2)}/\tau$, with τ being the FWHM of the pulse, ω and T_0 are the frequency and time period of the laser field, respectively, and φ is the CEP. In the calculation, the amplitude is set as 1 V/m and $\varphi = \pi/3$. The FWHM duration of the laser pulse is one optical cycle, and the carrier frequency of the laser field is 4.7×10^{15} Hz, which corresponds to 0.11388 atomic units (a.u.). As shown in Figure 1d, we clearly observe five LSPR peaks ordered as peaks 1–5 for this 90 nm silver cube, and they are located at

275 nm, 374 nm, 447 nm, 495 nm and 643 nm, respectively. These peaks correspond to five kinds of plasmon modes. The dominant peak is peak 4, which is located at 495 nm. The other peaks are much weaker than peak 4. Therefore, the overall extinction spectrum curve looks like a broad peak centered at 495 nm with three shoulders appearing on the short-wavelength side and one shoulder appearing on the long-wavelength side. Here, we try to assign unambiguously each LSPR peak to the excitation of either a dipole, or quadrupole, or hybrid resonance mode. Furthermore, to better understand the intrinsic nature of these plasmon resonance peaks, we analyze the physical origin of different LSPR peaks essentially from the surface charge distribution.

4. Discussion

The local field distribution has been widely utilized to discern the physical nature of a particular plasmon resonance mode appearing in the extinction spectrum curve. In fact, the optical properties of a particle are associated with surface modes, which are accompanied by polarization charge on the surface [11]. In other words, the optical extinction of the particle can be associated with the surface polarization charge. To understand clearly the physical origin of the surface modes, we calculate the electric field intensity patterns and the surface charge distributions for a silver particle under different incident waves. In the field intensity calculations, we have selected an enclosed cubic surface with a 2 nm gap from the surface of the nanocube. Figure 2 shows the calculated electric field intensity (in units of the incident field intensity throughout this paper) and charge density distribution (in units of C/m^3 throughout this paper) of the 90 nm silver nanocube in the xz plane with $y = -45$ nm at the 275 nm, 374 nm, 447 nm, 495 nm and 635 nm resonance peaks, respectively. Figure 2a shows the local electric field intensity distribution at the wavelength of 275 nm, whereas Figure 2b illustrates the corresponding polarized electric charge distribution. From the figures, we can clearly see that mode 1 is a dipole mode because charges of different sign are separated on the left and right parts of the face of the cube. For mode 2 and mode 3, the charge distribution shown in Figure 2d,f illustrates that both the 374 nm and 447 nm resonance modes are quadrupole modes, as the signs of the electric charge on the two z-axis edges are opposite to each other, and they are also opposite on the bottom side and upper side on one z-axis edge.

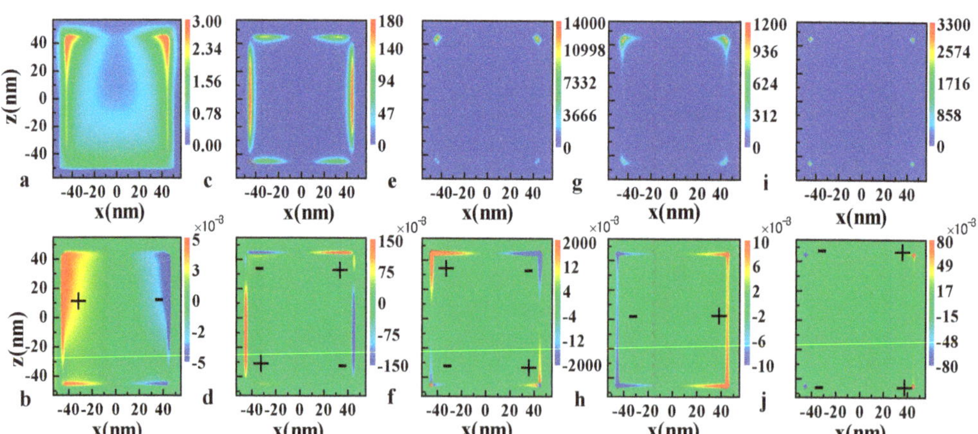

Figure 2. Calculated near-field intensity in the xz plane of a 90 nm silver nanocube at the (**a**) 275 nm, (**c**) 374 nm, (**e**) 447 nm, (**g**) 495 nm and (**i**) 643 nm resonance peaks, respectively. The corresponding polarized electric charge distribution at (**b**) 275 nm, (**d**) 374 nm, (**f**) 447 nm, (**h**) 495 nm and (**j**) 643 nm resonance peaks, respectively. Red and blue colors indicate positive and negative charge, which are also indicated by "+" and "−" marks, respectively.

According to Figure 2c,e, the field intensity is high at the edge of the cube for the 374 nm resonance mode, but for the 447 nm resonance mode the intensity is more concentrated in the corners. For mode 1 and mode 2, the polarization charges cover the face and the edges of the cube, respectively. However, polarization charges tend to concentrate in the corners when the resonance wavelength redshifts. Taking a closer look at Figure 2e,f, we find that both the local field and the charge distribution are larger on the upper side of the nanocube than on the bottom side. The difference is attributed to the retardation effect of the incident light, which becomes significant when the size of the particle increases to 90 nm. On the other hand, the charges in the two sharp corners at the bottom side are not strictly opposite in sign when compared with the upper side. Therefore, the excited plasmon mode at this wavelength is not a pure quadrupole mode. It can be assumed to be the superposition of dipole and quadrupole modes, and the quadrupole mode is dominant here. Figure 2g,h clearly show that the 495 nm mode corresponds to a dipole mode induced by the x axis polarized incident waves as the electric charge located on the edges and corners has opposite sign on the two z-axis edges in the xz plane with $y = -45$ nm. The intensity is concentrated in a larger volume near the upper two corners. Finally, mode 5, namely, the 635 nm resonance mode, is also a dipole mode and the charge is highly localized in the region that is very close to the corners as shown in Figure 2i,j.

From the discussion above, we can find several interesting features. First, the charge distribution follows a certain rule. Generally, the charges distributed on the faces of a cube contribute to the short-wavelength resonance mode while charges localized at the corners correspond to the long-wavelength surface mode. The mode generated by the edges of a cube locates between the two modes, and some modes originate from the combination of the faces, edges and corners. According to Ref. [11], the polarization charges on different parts of a nanocube produce different normal components of the polarization at the surface and result in the difference of dipole moment, which directly determine the surface mode frequencies. The relationship between the resonance frequencies and the charge distribution can be written in the form:

$$\langle \chi(\omega) \rangle = \sum \frac{C_m}{\chi^{-1}(\omega) + 4\pi n_m}, \tag{2}$$

where $\chi(\omega)$ is the complex dielectric susceptibility of the particle, n_m is the depolarization factor associated with the mth normal mode and C_m denotes the strength of the mode. A resonant peak occurs when $Re[\chi^{-1}(\omega)] \approx -4\pi n_m$ or $\varepsilon'(\omega) = 1 - n_m^{-1}$, where $\varepsilon'(\omega)$ is the value of the real part of the material dielectric function at each resonant peak. The depolarization factor decreases when the charge concentrates from the face to the corners of the cubic particle. As a result, the value of the real part of the material dielectric constant requested to support plasmon resonance decreases, and this corresponds to the redshift of the resonant wavelength. Second, as the particle is as large as 90 nm, quadrupole modes are excited by the incident wave at the metal nanoparticle due to the phase retardation of the field inside the particle. The 275 nm, 495 nm and 643 nm peaks are induced by the dipole resonances, whereas the 374 nm and 447 nm resonance modes are mainly quadrupole modes. The difference in the charge distribution and the contributions of the dipole component and quadrupole component reveal the physical origin of the five plasmon modes.

Third, the field enhancement factor does not scale proportionally to the extinction efficiency. Figure 1b shows that the dominant peak of the extinction cross section of a 90 nm Ag nanocube is 495 nm resonance mode while the 447 nm resonance mode is a shoulder of the dominant mode. Whereas the field intensity calculated in Figure 2e,g indicates that at a wavelength of 447 nm, the maximum field enhancement factor is 14,590, while it is only 1200 at a wavelength of 495 nm. Therefore, the 447 nm mode is a large field enhancement and low extinction (LFE-LE) resonance mode, according to the analysis made by Zhou and coworkers [35]. The physical mechanism behind this peculiar feature is that the polarization charge is highly densified in a very limited volume around the corners of the nanocube. As shown in Figure 2e,g, the field enhancement effect is highly localized in the region that is very close to the corner for mode 3. While for mode 4, the field enhancement effect takes place in a larger volume near the corner. The LFE-LE resonance mode may be very useful for practical applications such as SERS and nonlinear optical enhancement.

Until now, we have figured out the physical origin of the five surface plasmon modes and attributed each mode to a particular class of plasmonic excitation mode. Next, we provide an effective way to manipulate the optical properties of silver nanoparticles which are important for practical applications. As CEP is a very crucial parameter which can dramatically affect almost all dynamical processes in laser–matter interaction, the ultrashort pulse provides a new scheme to regulate and control the LSPRs from the aspect of excitation optical field engineering. Figure 3 shows the influence of CEP on the charge distribution for a quadrupole mode at 447 nm and a dipole mode at 643 nm. The sign of the surface charge alternates between negative and positive as CEP is changed from $\varphi = 0.25\pi$ to $\varphi = 1.25\pi$. When $\varphi = 0.25\pi$ and $\varphi = 1.25\pi$, the silver nanocube reaches a maximum charge polarization for two modes. The response of the conduction electron cloud to the incident electric field is closely related with the optical field retardation effects. As CEP can dramatically affect the surface charge distribution, which directly determines the optical properties of silver nanoparticles, CEP becomes a powerful modulation tool on the LSPRs modes. This can be clearly seen from the calculation result as shown in Figure 4. When $\varphi = 0.75\pi$, only three plasmon modes are excited, which locate at 275 nm, 374 nm and 495 nm, respectively. The extinction cross sections of these three modes are much higher than those with $\varphi = 0.25\pi$. However, the mode at 643 nm is much weaker and the peculiar LFE-LE resonance mode at 447 nm is drastically suppressed. The reason is that the nanocube reaches a minimum charge polarization for two modes when $\varphi = 0.75\pi$ as shown in Figure 3. The suppression or promotion of the LFE-LE resonance mode can find potential applications in the area of SERS and nonlinear optical enhancement. Based on our discussion above, CEP can act as a powerful modulation means for the regulation and control of LSPRs, e.g., to implement selective plasmon mode excitation.

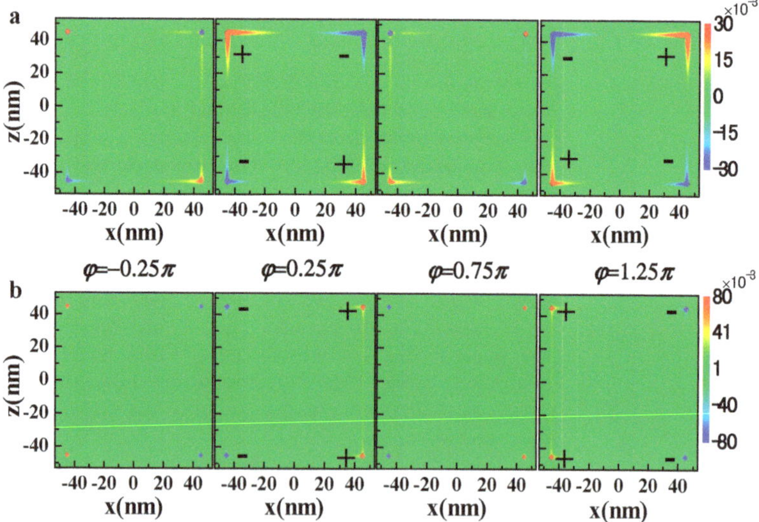

Figure 3. The influence of CEP on the surface charge distribution of a 90 nm silver nanocube. (**a**) Quadrupole mode at λ= 447 nm and (**b**) Dipole mode at $\lambda = 643$ nm with $\varphi = -0.25\pi$, 0.25π, 0.75π and 1.25π.

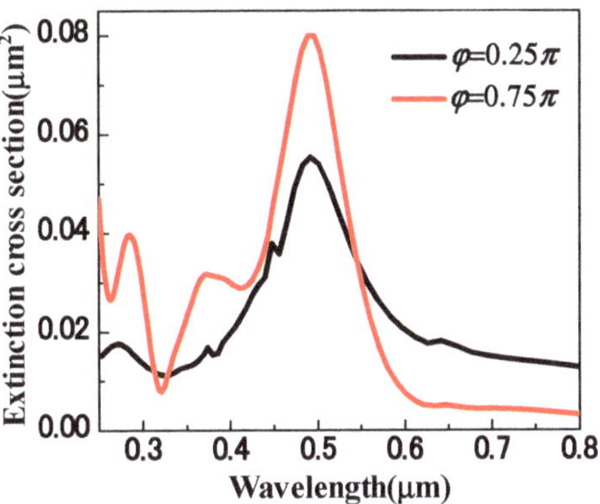

Figure 4. Comparison of the extinction cross section spectra for a 90 nm nanocubic silver particle with $\varphi = 0.25\pi$ and $\varphi = 0.75\pi$.

5. Conclusions

In summary, we have theoretically investigated the extinction spectra of a 90 nm silver nanocubic particle driven by a CEP-locked ultrashort pulse and found that five LSPR modes are excited by the incident waves. We have analyzed the physical origin of each mode from two aspects: the charge distribution on different parts of the cubic particle and the contributions of the dipole component and quadrupole component. The results show that the short-wavelength resonance mode mainly originates from the charges distributed at the faces of the cube, while the long-wavelength resonance mode mainly derives from the charges localized at the shape's corners. Charges located at edges contribute to the intermediate resonance modes located between the face mode and the corner mode, and other modes are generated by the cooperative contribution of charges at faces, edges and corners, together. On the other hand, the quadrupole mode occurs because of the onset of the electromagnetic retardation effect. From the two aspects, we have been able to specify unambiguously the physical origin of each LSPR peak in the extinction spectra.

We have presented an effective way to regulate and control the optical properties of nanoparticles by changing the CEP of the ultrashort excitation pulse. We have found that some plasmon modes can be drastically suppressed by changing CEP, especially the LFE-LE resonance mode. This suggests that CEP can act as a novel modulator of LSPR modes. As these LSPR modes are associated with very different optical properties regarding absorption, scattering, modal profile, local field enhancement, "hot spot" size and position and so on, the CEP could become a useful physical parameter to manipulate these properties for optimized applications in optical sensing, single-molecule detection, Raman spectroscopy, nonlinear optics, biomedical therapy and so on.

Author Contributions: Conceptualization, J.L. and Z.L.; methodology, Z.L.; software, J.L. and Z.L.; validation, J.L. and Z.L.; formal analysis, J.L. and Z.L.; investigation, J.L. and Z.L.; resources, Z.L.; data curation, J.L.; writing—original draft preparation, J.L.; writing—review and editing, Z.L.; visualization, J.L. and Z.L.; supervision, Z.L.; project administration, Z.L.; funding acquisition, J.L. and Z.L. All authors have read and agreed to the published version of the manuscript.

Funding: This research was funded by The National Natural Science Foundation of China, grant number 11904397.

Institutional Review Board Statement: Not applicable.

Informed Consent Statement: Not applicable.

Data Availability Statement: Not applicable.

Conflicts of Interest: The authors declare no conflict of interest.

References

1. Zhou, F.; Li, Z.; Liu, Y.; Xia, Y. Quantitative Analysis of Dipole and Quadrupole Excitation in the Surface Plasmon Resonance of Metal Nanoparticles. *J. Phys. Chem. C* **2008**, *112*, 20233–20240. [CrossRef]
2. Hao, E.; Schatz, G.C. Electromagnetic fields around silver nanoparticles and dimers. *J. Chem. Phys.* **2004**, *120*, 357–366. [CrossRef]
3. Kelly, K.L.; Coronado, E.; Zhao, L.L.; Schatz, G.C. The Optical Properties of Metal Nanoparticles: The Influence of Size, Shape, and Dielectric Environment. *J. Phys. Chem. B* **2003**, *107*, 668–677. [CrossRef]
4. Albrecht, M.G.; Creighton, J.A. Anomalously intense Raman spectra of pyridine at a silver electrode. *J. Am. Chem. Soc.* **1977**, *99*, 5215–5217. [CrossRef]
5. Liu, J.; Zheng, M.; Xiong, Z.; Li, Z. 3D dynamic motion of a dielectric micro-sphere within optical tweezers. *Opto-Electron Adv.* **2021**, *4*, 200015. [CrossRef]
6. Zhong, X.; Li, Z. Giant Enhancement of Near-Ultraviolet Light Absorption by TiO_2 via a Three-Dimensional Aluminum Plasmonic Nano Funnel-Antenna. *J. Phys. Chem. C* **2012**, *116*, 21547–21555. [CrossRef]
7. Liu, J.; Zhong, X.; Li, Z. Enhanced light absorption of silicon in the near-infrared band by designed gold nanostructures. *Chin. Phys. B* **2014**, *23*, 047306. [CrossRef]
8. Kumbhar, A.S.; Kinnan, M.K.; Chumanov, G. Multipole Plasmon Resonances of Submicron Silver Particles. *J. Am. Chem. Soc.* **2005**, *127*, 12444–12445. [CrossRef]
9. Sosa, I.O.; Noguez, C.; Barrera, R.G. Optical Properties of Metal Nanoparticles with Arbitrary Shapes. *J. Phys. Chem. B* **2003**, *107*, 6269–6275. [CrossRef]
10. Evanoff, D.D.; Chumanov, G. Size-controlled synthesis of nanoparticles. 1. "Silver-only" aqueous suspensions via hydrogen reduction. *J. Phys. Chem. B* **2004**, *108*, 13948–13956. [CrossRef]
11. Fuchs, R. Theory of the optical properties of ionic crystal cubes. *Phys. Rev. B* **1975**, *11*, 1732–1739. [CrossRef]
12. Mazzucco, S.; Geuquet, N.; Ye, J.; Stéphan, O.; Van Roy, W.; Van Dorpe, P.; Henrard, L.; Kociak, M. Ultralocal Modification of Surface Plasmons Properties in Silver Nanocubes. *Nano Lett.* **2012**, *12*, 1288–1294. [CrossRef]
13. Willets, K.A.; Van Duyne, R.P. Localized Surface Plasmon Resonance Spectroscopy and Sensing. *Annu. Rev. Phys. Chem.* **2007**, *58*, 267–297. [CrossRef] [PubMed]
14. Maier, S.A.; Atwater, H.A. Plasmonics: Localization and guiding of electromagnetic energy in metal/dielectric structures. *J. Appl. Phys.* **2005**, *98*, 011101. [CrossRef]
15. Nisoli, M.; De Silvestri, S.; Svelto, O.; Szipöcs, R.; Ferencz, K.; Spielmann, C.; Sartania, S.; Krausz, F. Compression of high-energy laser pulses below 5 fs. *Opt. Lett.* **1997**, *22*, 522–524. [CrossRef]
16. Steinmeyer, G.; Sutter, D.H.; Gallmann, L.; Matuschek, N.; Keller, U. Frontiers in Ultrashort Pulse Generation: Pushing the Limits in Linear and Nonlinear Optics. *Science* **1999**, *286*, 1507–1512. [CrossRef] [PubMed]
17. Yuan, K.; Bandrauk, A.D. Attosecond-magnetic-field-pulse generation by intense few-cycle circularly polarized UV laser pulses. *Phys. Rev. A* **2013**, *88*, 013417. [CrossRef]
18. Asaki, M.T.; Huang, C.; Garvey, D.; Zhou, J.P.; Kapteyn, H.C.; Murnane, M.M. Generation of 11-fs pulses from a self-mode-locked Ti:sapphire laser. *Opt. Lett.* **1993**, *18*, 977–979. [CrossRef]
19. Drexler, W.; Morgner, U.; Kartner, F.X.; Pitris, C.; Boppart, S.A.; Li, X.D.; Ippen, E.P.; Fujimoto, J.G. In vivo ultra-high-resolution optical coherence tomography. *Opt. Lett.* **1999**, *24*, 1221–1223. [CrossRef]
20. Sutter, D.H.; Steinmeyer, G.; Gallmann, L.; Matuschek, N.; Morier-Genoud, F.; Keller, U.; Scheuer, V.; Angelow, G.; Tschudi, T. Semiconductor saturable-absorber mirror–assisted Kerr-lens mode-locked Ti:sapphire laser producing pulses in the two-cycle regime. *Opt. Lett.* **1999**, *24*, 631–633. [CrossRef]
21. Brabec, T.; Krausz, F. Intense few-cycle laser fields: Frontiers of nonlinear optics. *Rev. Mod. Phys.* **2000**, *72*, 545–591. [CrossRef]
22. Lemell, C.; Tong, X.; Krausz, F.; Burgdörfer, J. Electron Emission from Metal Surfaces by Ultrashort Pulses: Determination of the Carrier-Envelope Phase. *Phys. Rev. Lett.* **2003**, *90*, 076403. [CrossRef]
23. Lenzner, M.; Krüger, J.; Sartania, S.; Cheng, Z.; Spielmann, C.; Mourou, G.; Kautek, W.; Krausz, F. Femtosecond Optical Breakdown in Dielectrics. *Phys. Rev. Lett.* **1998**, *80*, 4076–4079. [CrossRef]
24. Spielmann, C.; Burnett, N.H.; Sartania, S.; Koppitsch, R.; Schnürer, M.; Kan, C.; Lenzner, M.; Wobrauschek, P.; Krausz, F. Generation of Coherent X-rays in the Water Window Using 5-Femtosecond Laser Pulses. *Science* **1997**, *278*, 661–664. [CrossRef]
25. Hentschel, M.; Kienberger, R.; Spielmann, C.; Reider, G.A.; Milosevic, N.; Brabec, T.; Corkum, P.; Heinzmann, U.; Drescher, M.; Krausz, F. Attosecond metrology. *Nature* **2001**, *414*, 509–513. [CrossRef]
26. Wang, B.; Chen, J.; Liu, J.; Yan, Z.; Fu, P. Attosecond-pulse-controlled high-order harmonic generation in ultrashort laser fields. *Phys. Rev. A* **2008**, *78*, 023413. [CrossRef]
27. Cavalieri, A.L.; Goulielmakis, E.; Horvath, B.; Helml, W.; Schultze, M.; Fieß, M.; Pervak, V.; Veisz, L.; Yakovlev, V.S.; Uiberacker, M.; et al. Intense 1.5-cycle near infrared laser waveforms and their use for the generation of ultra-broadband soft-x-ray harmonic continua. *New J. Phys.* **2007**, *9*, 242. [CrossRef]
28. Liu, J.; Li, J.; Zhang, C.; Li, Z. Population trapping of a two-level atom via interaction with CEP-locked laser pulse. *J. Phys. Commun.* **2018**, *2*, 085017. [CrossRef]
29. Goulielmakis, E.; Schultze, M.; Hofstetter, M.; Yakovlev, V.S.; Gagnon, J.; Uiberacker, M.; Aquila, A.L.; Gullikson, E.M.; Attwood, D.T.; Kienberger, R.; et al. Single-cycle nonlinear optics. *Science* **2008**, *320*, 1614–1617. [CrossRef] [PubMed]

30. Miller, M.M.; Lazarides, A.A. Sensitivity of metal nanoparticle surface plasmon resonance to the dielectric environment. *J. Phys. Chem. B* **2005**, *109*, 21556–21565. [CrossRef]
31. Mulvaney, P. Surface plasmon spectroscopy of nanosized metal particles. *Langmuir* **1996**, *12*, 788–800. [CrossRef]
32. Yee, K.S. Numerical Solution of Initial Boundary Value Problems Involving Maxwells Equations in Isotropic Media. *IEEE Trans. Antennas Propag.* **1966**, *14*, 302–307.
33. Taflove, A.; Hagness, S.C. *Computational Electrodynamics: The Finite-Difference Time-Domain Method*, 3rd ed.; Artech House: Norwood, MA, USA, 2005; pp. 475–506.
34. Palik, E.D. *Handbook of Optical Constants of Solids*; Academic Press: Orlando, FL, USA, 1985; pp. 547–569.
35. Zhou, F.; Liu, Y.; Li, Z. Simultaneous low extinction and high local field enhancement in Ag nanocubes. *Chin. Phys. B* **2011**, *20*, 037303. [CrossRef]

Article

Key Distribution Scheme for Optical Fiber Channel Based on SNR Feature Measurement

Xiangqing Wang [1], Jie Zhang [1,*], Bo Wang [1], Kongni Zhu [1], Haokun Song [1], Ruixia Li [2] and Fenghui Zhang [2]

[1] State Key Laboratory of Information Photonics and Optical Communications, Beijing University of Posts and Telecommunications, Beijing 100876, China; wxqing@bupt.edu.cn (X.W.); wb1059@bupt.edu.cn (B.W.); zhukongni@bupt.edu.cn (K.Z.); haokunsong@bupt.edu.cn (H.S.)

[2] College of Electronics and Information Engineering, West Anhui University, Lu'an 237000, China; 03000058@wxc.edu.cn (R.L.); zhangfenghui@wxc.edu.cn (F.Z.)

* Correspondence: jie.zhang@bupt.edu.cn; Tel.: +86-139-1106-0930

Abstract: With the increase in the popularity of cloud computing and big data applications, the amount of sensitive data transmitted through optical networks has increased dramatically. Furthermore, optical transmission systems face various security risks at the physical level. We propose a novel key distribution scheme based on signal-to-noise ratio (SNR) measurements to extract the fingerprint of the fiber channel and improve the physical level of security. The SNR varies with time because the fiber channel is affected by many physical characteristics, such as dispersion, polarization, scattering, and amplifier noise. The extracted SNR of the optical fiber channel can be used as the basis of key generation. Alice and Bob can obtain channel characteristics by measuring the SNR of the optical fiber channel and generate the consistent key by quantization coding. The security and consistency of the key are guaranteed by the randomness and reciprocity of the channel. The simulation results show that the key generation rate (KGR) can reach 25 kbps, the key consistency rate (KCR) can reach 98% after key post-processing, and the error probability of Eve's key is ~50%. In the proposed scheme, the equipment used is simple and compatible with existing optic fiber links.

Keywords: key distribution; signal-to-noise ratio; reciprocity; key consistency rate

Citation: Wang, X.; Zhang, J.; Wang, B.; Zhu, K.; Song, H.; Li, R.; Zhang, F. Key Distribution Scheme for Optical Fiber Channel Based on SNR Feature Measurement. *Photonics* **2021**, *8*, 208. https://doi.org/10.3390/photonics8060208

Received: 6 May 2021
Accepted: 4 June 2021
Published: 9 June 2021

1. Introduction

With the sharp increase in the speed and distance of optical communications, optical networks have become more accessible; accordingly, more threats and higher risks may be posed. A reliable key distribution system is required to solve this problem and ensure communication security. Traditional public key-based key distribution security primarily depends on the complexity of an algorithm, such as the RSA algorithm [1,2]. However, traditional cryptography is susceptible to the rapid progress of hardware and algorithms. The robustness of these algorithms faces severe challenges with the development of computer systems, especially quantum computers [3,4]. Quantum key distribution (QKD) is theoretically considered as the only solution to guarantee absolute security at the physical level [5–8]. However, QKD requires highly sensitive optical detection equipment instead of optical amplifiers. Therefore, it is still challenging to realize QKD at longer distances and higher key rates.

A promising and cost-effective approach is to take advantage of the unpredictable and random characteristics of the transmission channel to convert the random characteristics of the environment into a secure key. In this approach, the key is only highly correlated to the legitimate users (Alice and Bob), but not to the signal being eavesdropped on by Eve. This idea has already been put into practice in wireless and fiber-optic communications systems. In a wireless communication system, for key generation, the random fading effects of the wireless channel are utilized [9,10], such as received signal strength, channel impulse response, and frequency phase in key distribution schemes [11,12].

Unlike wireless communication, the optic fiber links have stronger resistance to environmental disturbance. The proposed method achieves the implementation of a classical physical layer secure key distribution (SKGD) with unique fiber characteristics, which extracts the key from the fiber channel characteristics. For the key distribution scheme based on chaotic synchronization, the security has been improved [13,14], but it is incompatible with the existing optic fiber link configuration, and the key distribution distance is short. The SKGD scheme based on the unique characteristics of the physical layer has the advantages of high security, low cost, and a simple structure. The polarization mode dispersion (PMD) depends on the birefringence distribution in instantaneous space [15–18] and fluctuates randomly in the link, providing a random source for key generation. Eve close to the legitimate party can easily cause information disclosure because the PMD can only produce a favorable effect when the distance is long enough. Based on the key distribution scheme of the optical fiber interferometer [19,20], the interferometer is exposed to the public and vulnerable to active intrusion attacks. As a result, other solutions that can guarantee high security and easy implementation have not been fully explored.

Recently, researchers have proposed a key generation scheme based on the bit error rate (BER) measurement of physical channel characteristics [21,22]. The scheme takes advantage of the physical channel's randomness to ensure the security of the key and does not change the structure of the existing optic fiber link. However, this scheme requires a high BER in order for the normal transmission to be affected. We propose a key distribution measurement scheme based on the SNR characteristics of the physical layer of the optical fiber channel without affecting the normal transmission. The combination of key distribution and encryption transmission is realized by simulation. The final KGR can reach 25 kbps, and the KCR can reach 98%. Fingerprint SNR is used as the random source of key extraction in the system, and the uniqueness of the fiber channel ensures the high security of the generated key.

2. Key Generation Scheme for Optical Fiber Communication

2.1. Channel Model

As shown in Figure 1, the signal sent by the legitimate client Alice is A^n, while the legitimate receiver Bob receives the signal sequence B^n. The eavesdropping signal obtained by Eve is E^n. The legitimate receiver obtains the SNR by comparing the signal from the receiver with that from the transmitter. According to information theory, the mutual information between Eve and Alice as Equation (1) should be as small as possible for reliable transmission. Eve cannot correctly obtain Alice's key information, indicating that the key of communication negotiation is secure when $I(A^n; E^n) \to 0$.

$$I(A^n; E^n) = H(A^n) - H(A^n|E^n) \tag{1}$$

The keys generated by Alice, Bob, and Eve are $K_A = f_A(A^n)$, $K_B = f_A(B^n)$, and $K_E = f_E(E^n)$, respectively, to achieve the security of the physical layer. Assuming that the coefficient of key consistency ε is large enough for n to satisfy the following relation, the system is secure [9,23].

$$P(K_A = K_B) \geq \varepsilon \tag{2}$$

$$I(K_A; E^n) \leq 1 - \varepsilon \tag{3}$$

In the best case, the key consistency rate approached in Equation (2) indicates that the completion of the key generated by Alice and Bob is consistent. Equation (3) indicates that Eve receives the information irrespective of the key generated by the legitimate. The maximum security key capacity is shown in Equation (4). The key capacity is C_{AB} when the communication between legitimate parties is normal, and the generated key capacity is C_{AE} when Eve eavesdrops. The maximum value C_K is obtained by subtracting C_{AE} from C_{AB}.

$$C_K = [C_{AB} - C_{AE}] = \max[I(A^n; B^n) - I(A^n; E^n)] \tag{4}$$

$$C_K = [C_{AB} - C_{AE}] = [\frac{1}{2}\log(1 + \frac{p}{\sigma_1^2}) - \frac{1}{2}\log(1 + \frac{p}{\sigma_1^2 + \sigma_2^2})] \quad (5)$$

The security capacity $C_K > 0$ under the Gaussian Tap channel is shown in Equation (5), where p is the power of the signal, and $\sigma_i (i = 1,\ 2)$ are, respectively, the noise variance of Bob and Eve stealing channels in the normal transmission main channel.

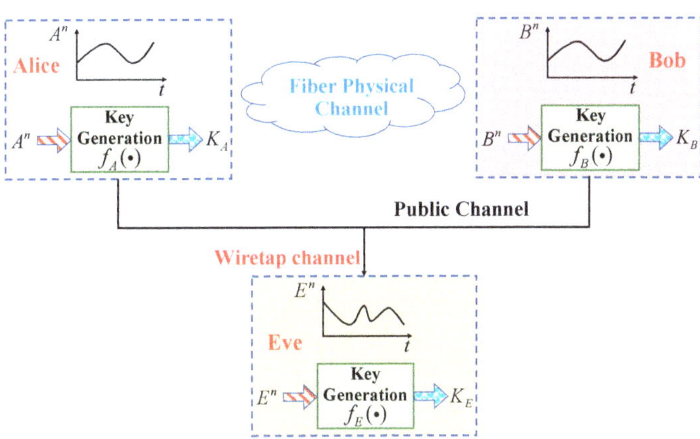

Figure 1. Key generation model diagram.

2.2. Channel Model

Figure 2 shows the proposed key generation scheme based on the SNR measurement in the optical fiber loop. Due to the changes in temperature, external stress, and physical parameters, the measured SNR fluctuates randomly. Specific physical layer channel features are extracted as follows: Alice and Bob simultaneously measure the SNR parameter changes in the fiber loop. Data from Alice and Bob are transmitted in the fiber loop link through time-division multiplexing. Alice and Bob occupy different time slots, i.e., even and odd symmetric time slots, respectively. First, Alice generates random data and random ground state C^A through a PRNG. Afterwards, the DSP module of the transmitter generates the signal D^{AB} from the signal D^A and the ground state C^A. The signal D^{AB} reaches the Bob terminal after transmission in the optical fiber. Bob's DSP module processes the received signal with a random ground state C^B to generate the data D^{AB}. After the data D^{AB} signal is transmitted in the optical fiber, the signal N^{AB} with channel noise reaches Alice. The receiver's DSP module generates data D^{ABA} based on the received signal D^{AB} and random ground state C^A. The SNR performance of the fiber loop can be obtained by comparing the data D^A with D^{ABA}. Similarly, Bob can measure the fiber loop simultaneously to acquire the SNR change rate of the fiber loop backlink.

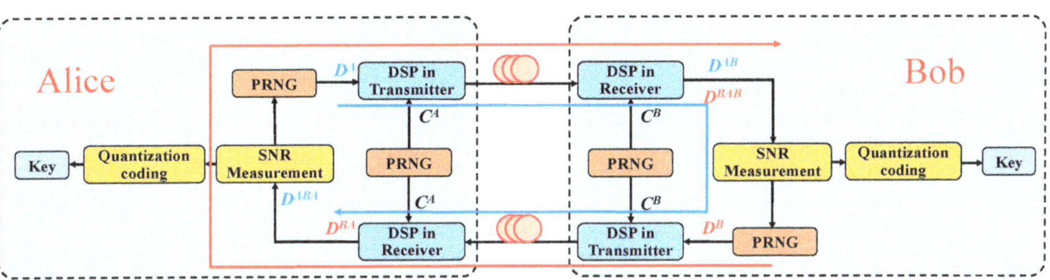

Figure 2. Schematic diagram of key generation principle at the physical layer. PRNG: pseudo-random number generator. DSP: digital signal processing.

2.3. Key Evaluation Index

In this paper, the SNR as the channel physical layer feature is introduced to evaluate the system performance. The calculation of the SNR is shown as follows: The bit resolution of signal quantization is N, the quantization noise variance is σ_e^2, and the signal's power is σ_s^2. The expression of SNR is shown in Equation (6) [24].

$$SNR = 10\log_{10}\frac{\sigma_s^2}{\sigma_e^2} = 6.02N + 10.79 + 10\log_{10}\sigma_s^2(dB) \tag{6}$$

$$KEY = \begin{cases} 0 & if\ SNR < ave(SNR) * (1-\alpha_q) \\ 1 & if\ SNR > ave(SNR) * (1+\alpha_q) \end{cases} \tag{7}$$

Alice and Bob obtain the SNR through the loopback measurement and then quantize it to generate the key. As presented in Equation (7), *ave* is the average value of the SNR, and α_q is the quantization coefficient.

KGR is the number of keys generated in a period divided by the test time. The details are shown in Equation (8).

$$KGR = \frac{R}{\ell_N}(1-\alpha_q)(1-\xi) \tag{8}$$

R is the signal transmission rate of 10 Gb/s, and ℓ_N is the segment length of the received signal. Moreover, α_q denotes the outliers during quantization corresponding to the number of key bits generated after each symbol is quantized as 0.3, and ξ is the discarding rate of privacy amplification (i.e., bit rate discarded in privacy amplification).

$$KCR = 1 - \frac{\sum_{j}^{\ell_N}|K_{Alice}(j) - K_{Bob}(j)|}{\ell_N} \tag{9}$$

The KCR is the consistency rate of the key generated by Alice and Bob, and the partition length of the key is ℓ_N, as shown in Equation (9).

$$KER = \frac{\sum_{j}^{\ell_N}|K_{Alice}(j) - K_{Eve}(j)|}{\ell_N} \tag{10}$$

The key error rate (KER) is the error rate of the key generated by Alice and Eve, as shown in Equation (10).

2.4. Key Generation Process

The key quantization optimization process is shown in Figure 3. Alice and Bob extract channel characteristics of the SNR, quantify and code to generate the key, and generate the key with excellent consistency and security after post-processing. The specific steps are as follows:

Step 1: Bob calculates the channel's SNR through loopback measurements;

Step 2: The characteristic information SNR is quantized to generate a consistency key by encoding. The specific steps are shown in Equation (2);

Step 3: The key's consistency is judged, and the consistency factor is set as $\varepsilon = 0.95$. If $KCR > \varepsilon$, it will proceed to the next step; otherwise, the quantization factor α_q will be updated and it will go back to the beginning;

Step 4: After the key is post-processed, Bob transmits the quantization coefficient α_q and consistency factor ε to Alice through the public channel;

Step 5: To increase the security of the generated key, data with length 0 or length 1 are discarded, and the key sequence KB with good randomness and consistency is obtained;

Step 6: Bob shares the optimized parameters α_q and ε with Alice, and Alice generates the key sequence KA in the same way.

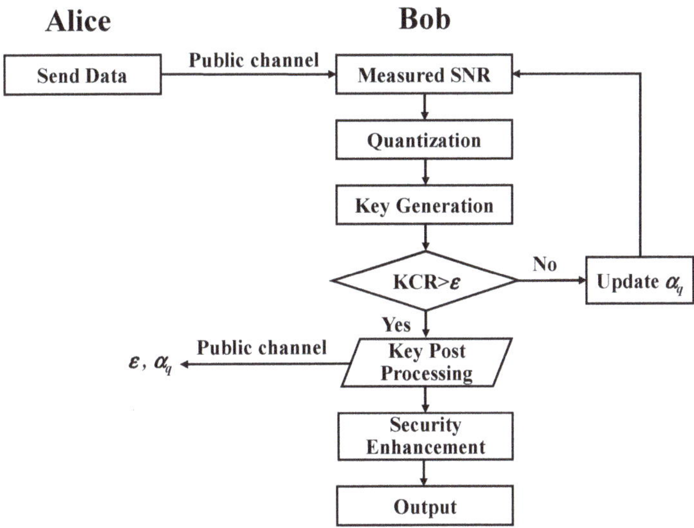

Figure 3. Optimized key generation factors ε and α_q.

3. Key Distribution Simulation Platform Setup

As shown in Figure 4, the data transmission rate of the scheme is 10 Gbps, the optic fiber link is 200 km, the laser transmitting power is 1 mW, and the wavelength is 1550 nm. The sending process of the scheme is as follows: Firstly, Alice obtains the orthogonal frequency division multiplexing signal encrypted by quantum noise stream through DSP processing. Next, the signal is converted from digital to analogue through the arbitrary waveform generator (AWG), and the electrical signal is modulated onto an optical carrier using an I/Q modulator. The signal amplified by erbium-doped fiber amplifier (EDFA) enters the optic fiber link to Bob's terminal. It is demodulated by the coherent receiver and sampled by a 20 GSa/s oscilloscope (OSC). Bob converts the signal to analogue through AWG and then modulates it to the optical carrier using the I/Q modulator. After passing through the amplifier and optic fiber link, Alice's coherent receiver carries out coherent demodulation on the signal. The demodulated data are sent to an OSC for sampling (sampling rate = 20 GSa/s). Finally, Alice performs DSP processing on the sampled signal. There are simulation-specific parameters, as listed in Table 1.

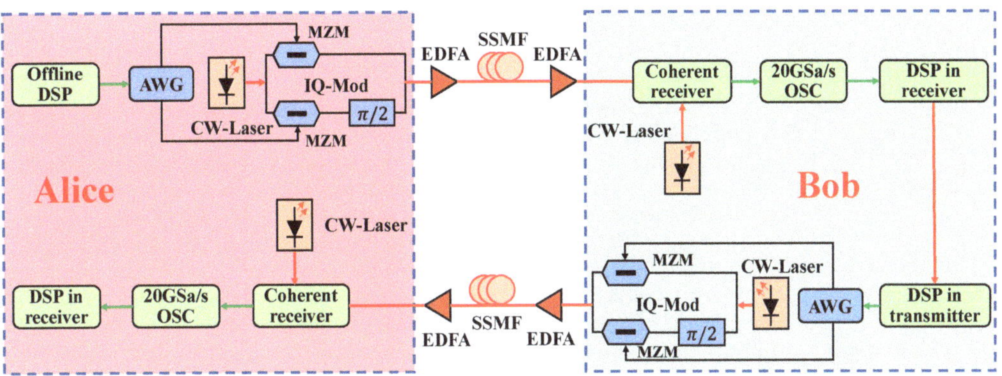

Figure 4. System simulation platform. MZM: Mach–Zehnder modulator. SSMF: standard single mode fiber.

Table 1. Simulation parameter list.

Equipment	Parameter Configuration
AWG	Transmitting Rate: 10 Gb/s
Light source	Wavelength: 1550 nm
EDFA	Launch Power: 1 mW
Ultra-low loss fiber	Power: 12 dBm
OSC	200 km, 0.2 dBm/km

4. Analysis of Simulation Results

4.1. Random Analysis

The National Institute of Standards and Technology (NIST) random test results of key sequences are illustrated in Figure 5a to evaluate the randomness of key bits. We randomly selected a set of keys from the key sequence as the key test sequence, and 10 NIST sub-tests [25]. The return threshold of all 10 of the NIST sub-tests is above 0.99, which indicates that the key sequence has true randomness. In addition, for some tests that return multiple thresholds, we only give the minimum value. Figure 5b describes the 0/1 ratio performance of Alice and Bob's key bits as a function of the SNR sampling interval. Alice quantifies the key to receive the 0, 1 key. The noise sources, such as laser, optical amplifier, and physical fiber parameters, determine the randomness of the fiber channel. Consequently, the probability of keys 0 and 1 fluctuates around 0.5, further verifying the key's randomness.

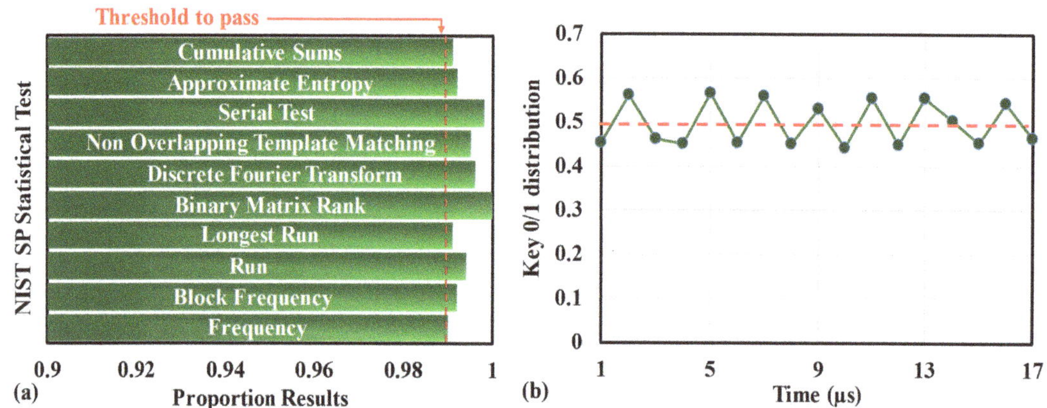

Figure 5. (**a**) Key NIST test results, and (**b**) 0/1 distribution probability of the key.

4.2. Consistency Analysis

Key consistency comes from precise channel reciprocity, so Alice and Bob share a highly related key source. According to the physical characteristics of the reciprocity of the channel, a consensus key is generated. The reciprocity of the channel mainly improves the consistency of the key. The metric characteristic of the channel generated by the key is the SNR. As shown in Figure 6a, the KGR of the system first increases and then decreases. With the increase in the SNR sampling interval, the calculation of BER requires more data bits, leading to the decrease in the KGR. When the sampling time is 2 μs, the maximum KGR is 25 kbps. The amount of data obtained at each sampling point is relatively insignificant when the sampling interval is very small, leading to a relatively severe miscalculation of SNR. Therefore, the SNR values calculated by Alice and Bob are inconsistent, which affects the KGR. When the sampling interval is large, the SNR value calculated within the same amount of time is relatively small, leading to a relatively low KGR rate. Therefore, it is necessary to set a reasonable sampling interval to obtain a relatively high KGR. It is obvious that faster fluctuation of channels corresponds to a smaller sampling interval, thus extracting more channel feature and generating more secret keys.

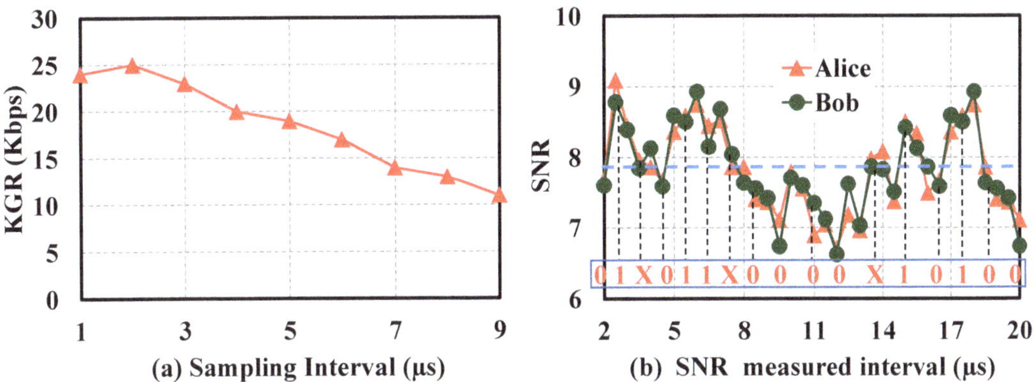

Figure 6. (**a**) KGR change curve and (**b**) key extraction quantization curve.

In Figure 6b, red represents the sampling value measured by Alice, and green characterizes the sampling value measured by Bob. It is evident that the change curves of Alice and Bob almost coincide, and the curve consistency is significant since they simultaneously measure the same fiber. As a result, we quantize the sample value to one and zero when it is greater and less than the mean, respectively. In addition, if the sample value is near the mean, we discard the data.

Furthermore, the KGR changes when varying the transmission power, sampling interval, and system's laser power. The signal transmission rate is altered to extract more channel features, and the KGR increases when the transmission rate is relatively low. However, the KGR does not increase and tends to be stable when the transmission rate increases to a certain extent. The sampling interval is varied to find the maximum KGR from different sampling intervals. At this time, the change in the KGR should be non-linear to determine the suitable KGR.

As shown in Figure 7, the relationship between the length of the selected data and the consistency is direct. The smaller the data block length is, the worse the key consistency is. By quantifying the SNR, the consistency of the initial keys obtained by Alice and Bob is as above. The SNR interval gradually increases with key consistency, and the obtained key consistency rate reaches 98% when the SNR interval is greater than 1.4 μs. As the SNR measurement interval increases, the error curve is smoother, and the fluctuation reduces, so the key consistency quantized by Alice and Bob is enhanced. The extracted keys can be subjected to off-line DSP to eliminate the influence of the inconsistency between the two key sequences of Alice and Bob.

More extensive data are needed to calculate the SNR required to increase the KCR at both ends, because the longer the data, the more stable the SNR and the higher the KCR. If the selected data are too long, the KGR for generating the key decreases, in that the calculated SNR drops for extensive data. Therefore, a reasonable SNR measurement interval should be selected to strike a balance between the KCR and the KGR.

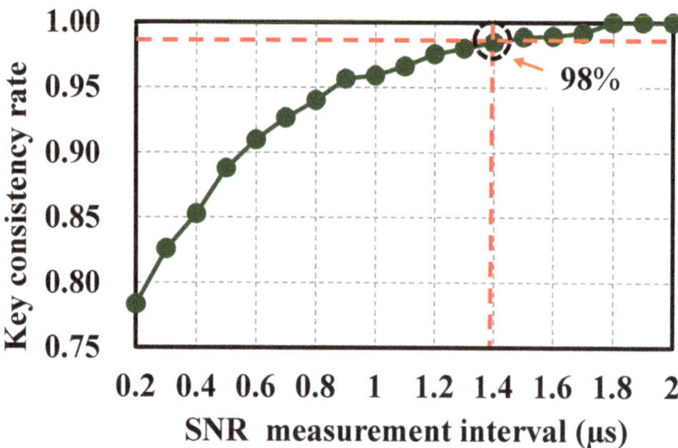

Figure 7. Influence of SNR measurement interval on Alice–Bob consistency.

4.3. Security Analysis

As shown in Figure 8a, the measured SNR value for Alice and Bob is approximately 10, while Eve's SNR is around −30. Alice and Bob loop back the measurement, and they know the initial key to decrypt the signal. Eve can only intercept information from the optical fiber without the initial key, so the SNR is low.

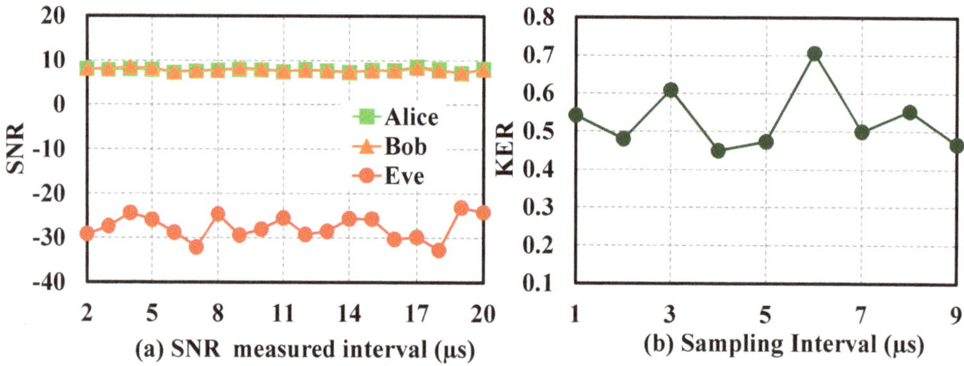

Figure 8. (**a**) The measured value curve of Alice and Bob Eve SNR, and (**b**) the change curve of the error rate of Eve key.

As shown in Figure 8b, Eve quantifies the key for SNR and Alice, then she obtains the key for comparison. The error rate of receiving the key is about 50%, so Eve cannot obtain the correct key. Moreover, the correlation coefficient between Alice and Eve is $cc = 0.02$, indicating that the key generated by the system has superior security. Alice and Eve's SNR is not the same, since Eve measures the SNR through different equipment and lines from Alice and Bob. The SNR is a variable that can be used to negotiate keys to ensure the generated keys are unique.

5. Conclusions

In this paper, a key generation scheme based on the SNR characteristic of the optical fiber channel's physical layer is proposed. In the scheme, the rate of SNR change is obtained using loopback measurements, and the changed SNR is compared with the threshold system to judge whether the system is attacked and the legitimate equipment can be correctly authenticated. If Eve launches an active attack on the system, it impacts the security and reliability of the generated key. Therefore, we can use physical layer authentication for intrusion detection. The simulation results show that the KGR can reach 25 kbps, and the KCR can reach 98%. The correlation coefficient of BER measurement samples of Alice and Eve is relatively low. The correlation coefficient $cc = 0.02$ and the KER of Eve is only 50%, indicating that this system has high security. Due to the physical characteristics of using the SNR as a key distribution scheme, normal communication requirements can be guaranteed to set the SNR values. The scheme can be used together with other security methods in the higher network layer, to enhance communication security and withstand active intrusion attacks. It has excellent application value and is suitable for popularization.

Author Contributions: Conceptualization, X.W. and B.W.; methodology, X.W.; software, B.W., K.Z. and H.S.; validation, X.W., B.W. and K.Z.; formal analysis, H.S.; investigation, H.S.; resources, R.L.; data curation, R.L.; writing—original draft preparation, X.W. and F.Z.; writing—review and editing, F.Z.; visualization, F.Z.; supervision, J.Z.; project administration, X.W.; funding acquisition, J.Z. All authors have read and agreed to the published version of the manuscript.

Funding: This work is supported by the NSFC (Grant No.:61831003), University Natural Science Research Project of Anhui Province (Grant No.: KJ2019A0616), and the Key Projects in Natural Science of West Anhui University (Grant No.: WXZR201719). The authors are thankful for the funding support from the Anhui Provincial Quality Engineering Project (Grant No.: 2020jyxm2152).

Institutional Review Board Statement: Not applicable.

Informed Consent Statement: Not applicable.

Data Availability Statement: This study does not report any data.

Acknowledgments: We acknowledge the support given by Kai Wang and Shuang Wei during the project.

Conflicts of Interest: The authors declare no conflict of interest. The funders had no role in the design of the study; in the collection, analysis, or interpretation of data; in the writing of the manuscript, or in the decision to publish the results.

References

1. Rivest, R.L.; Shamir, A.; Adleman, L. A method for obtaining digital signatures and public-key cryptosystems. *Commun. ACM* **1978**, *21*, 120–126. [CrossRef]
2. Katz, J.; Lindell, Y. *Introduction to Modern Cryptography*; CRC Press: Boca Raton, FL, USA, 2020.
3. Shor, P. Algorithms for quantum computation: Discrete logarithms and factoring. In Proceedings of the 35th Annual Symposium on Foundations of Computer Science, Santa Fe, NM, USA, 20–22 November 1994; pp. 124–134.
4. Patra, B.; Incandela, R.M.; Van Dijk, J.P.G.; Homulle, H.A.R.; Song, L.; Shahmohammadi, M.; Staszewski, R.B.; Vladimirescu, A.; Babaie, M.; Sebastiano, F.; et al. Cryo-CMOS Circuits and Systems for Quantum Computing Applications. *IEEE J. Solid-State Circuits* **2017**, *53*, 309–321. [CrossRef]
5. Lo, H.-K.; Curty, M.; Tamaki, K. Secure quantum key distribution. *Nat. Photon.* **2014**, *8*, 595–604. [CrossRef]
6. Korzh, B.; Lim, C.C.W.; Houlmann, R.; Gisin, N.; Li, M.J.; Nolan, D.A.; Sanguinetti, B.; Thew, R.; Zbinden, H. Provably secure and practical quantum key distribution over 307 km of optical fibre. *Nat. Photon.* **2015**, *9*, 163–168. [CrossRef]
7. Lucamarini, M.; Yuan, Z.L.; Dynes, J.F.; Shields, A.J. Overcoming the rate–distance limit of quantum key distribution without quantum repeaters. *Nat. Cell Biol.* **2018**, *557*, 400–403. [CrossRef] [PubMed]
8. Hwang, W.-Y. Quantum Key Distribution with High Loss: Toward Global Secure Communication. *Phys. Rev. Lett.* **2003**, *91*, 057901. [CrossRef]
9. Zhang, J.; Duong, T.Q.; Marshall, A.; Woods, R. Key Generation from Wireless Channels: A Review. *IEEE Access* **2016**, *4*, 614–626. [CrossRef]
10. Bottarelli, M.; Epiphaniou, G.; Ben Ismail, D.K.; Karadimas, P.; Al-Khateeb, H. Physical characteristics of wireless communication channels for secret key establishment: A survey of the research. *Comput. Secur.* **2018**, *78*, 454–476. [CrossRef]
11. Premnath, S.N.; Jana, S.; Croft, J.; Gowda, P.L.; Clark, M.; Kasera, S.K.; Patwari, N.; Krishnamurthy, S. Secret Key Extraction from Wireless Signal Strength in Real Environments. *IEEE Trans. Mob. Comput.* **2013**, *12*, 917–930. [CrossRef]
12. Liu, Y.; Draper, S.C.; Sayeed, A.M. Exploiting Channel Diversity in Secret Key Generation from Multipath Fading Randomness. *IEEE Trans. Inf. Forensics Secur.* **2012**, *7*, 1484–1497. [CrossRef]
13. Sasaki, T.; Kakesu, I.; Mitsui, Y.; Rontani, D.; Uchida, A.; Sunada, S.; Yoshimura, K.; Inubushi, M. Common-signal-induced synchronization in photonic integrated circuits and its appli-cation to secure key distribution. *Opt. Express* **2017**, *25*, 26029–26044. [CrossRef] [PubMed]

14. Zhao, Z.; Cheng, M.; Luo, C.; Deng, L.; Zhang, M.; Fu, S.; Tang, M.; Shum, P.; Liu, D. Semiconductor-laser-based hybrid chaos source and its application in secure key distribution. *Opt. Lett.* **2019**, *44*, 2605–2608. [CrossRef] [PubMed]
15. Zhang, L.; Hajomer, A.A.E.; Yang, X.; Hu, W. Error-free secure key generation and distribution using dynamic Stokes parameters. *Opt. Express* **2019**, *27*, 29207–29216. [CrossRef]
16. Hajomer, A.A.E.; Zhang, L.; Yang, X.; Hu, W. Accelerated key generation and distribution using polarization scrambling in optical fiber. *Opt. Express* **2019**, *27*, 35761–35773. [CrossRef]
17. Hajomer, A.A.E.; Zhang, L.; Yang, X.; Hu, W. Post-Processing Protocol for Physical-Layer Key Generation and Distribution in Fiber Networks. *IEEE Photon. Technol. Lett.* **2020**, *32*, 901–904. [CrossRef]
18. Zaman, I.U.; Lopez, A.B.; Al Faruque, M.A.; Boyraz, O. Physical Layer Cryptographic Key Generation by Exploiting PMD of an Optical Fiber Link. *J. Light. Technol.* **2018**, *36*, 5903–5911. [CrossRef] [PubMed]
19. Huang, C.; Ma, P.Y.; Blow, E.C.; Mittal, P.; Prucnal, P.R. Accelerated secure key distribution based on localized and asymmetric fiber interfer-ometers. *Opt. Express* **2019**, *27*, 32096–32110. [CrossRef] [PubMed]
20. Kravtsov, K.; Wang, Z.; Trappe, W.; Prucnal, P.R. Physical layer secret key generation for fiber-optical networks. *Opt. Express* **2013**, *21*, 23756–23771. [CrossRef]
21. Wang, X.; Zhang, J.; Li, Y.; Zhao, Y.; Yang, X. Secure Key Distribution System Based on Optical Channel Physical Features. *IEEE Photon. J.* **2019**, *11*, 1–11. [CrossRef]
22. Lei, C.; Zhang, J.; Li, Y.; Zhao, Y.; Wang, B.; Gao, H.; Li, J.; Zhang, M. Long-haul and High-speed Key Distribution Based on One-way Non-dual Arbitrary Basis Transformation in Optical Fiber Link. In Proceedings of the Optical Fiber Communication Conference and Exhibition(OFC), San Diego, CA, USA, 8–12 March 2020; p. W2A.51.
23. Maurer, U.M. Secret key agreement by public discussion from common information. *IEEE Trans. Inf. Theory* **1993**, *39*, 733–742. [CrossRef]
24. Yang, X.; Zhang, J.; Li, Y.; Gao, G.; Zhao, Y.; Zhang, H. Single-carrier QAM/QNSC and PSK/QNSC transmission systems with bit-resolution limited DACs. *Opt. Commun.* **2019**, *445*, 29–35. [CrossRef]
25. Fratalocchi, A.; Fleming, A.; Conti, C.; Di Falco, A. NIST-certified secure key generation via deep learning of physical unclonable functions in silica aerogels. *Nanophotonics* **2020**, *10*, 457–464. [CrossRef]

 photonics

Communication

Dispersion Optimization of Silicon Nitride Waveguides for Efficient Four-Wave Mixing

Yaping Hong [1,†], Yixiao Hong [2,†], Jianxun Hong [1,*] and Guo-Wei Lu [3,†]

1 School of Information Engineering, Wuhan University of Technology, Wuhan 430070, China; hyping@whut.edu.cn
2 School of Electrical and Electronic Engineering, Hubei University of Technology, Wuhan 430068, China; yxhong@hbut.edu.cn
3 Division of Computer Engineering, The University of Aizu, Fukushima 965-8580, Japan; guoweilu@u-aizu.ac.jp
* Correspondence: jxhong@whut.edu.cn
† These authors contributed equally to this work.

Abstract: Silicon nitride waveguides have emerged as an excellent platform for photonic applications, including nonlinear optical signal processing, owing to their relatively high Kerr nonlinearity, negligible two photon absorption, and wide transparent bandwidth. In this paper, we propose an effective approach using 3D finite element method to optimize the dispersion characteristics of silicon nitride waveguides for four-wave mixing (FWM) applications. Numerical studies show that a flat and low dispersion profile can be achieved in a silicon nitride waveguide with the optimized dimensions. Near-zero dispersion of 1.16 ps/km/nm and 0.97 ps/km/nm at a wavelength of 1550 nm are obtained for plasma-enhanced chemical vapor deposition (PECVD) and low-pressure chemical vapor deposition (LPCVD) silicon nitride waveguides, respectively. The fabricated micro-ring resonator with the optimized dimensions exhibits near-zero dispersion of −0.04 to −0.1 ps/m/nm over a wavelength range of 130 nm which agrees with the numerical simulation results. FWM results show that near-zero phase mismatch and high conversion efficiencies larger than −12 dB using a low pump power of 0.5 W in a 13-cm long silicon nitride waveguide are achieved.

Keywords: four-wave mixing; dispersion; phase matching; silicon nitride waveguide; conversion efficiency

Citation: Hong, Y.; Hong, Y.; Hong, J.; Lu, G.-W. Dispersion Optimization of Silicon Nitride Waveguides for Efficient Four-Wave Mixing. *Photonics* **2021**, *8*, 161. https://doi.org/10.3390/photonics8050161

Received: 17 April 2021
Accepted: 10 May 2021
Published: 11 May 2021

1. Introduction

With the rapid development of the modern optical communication, the technology of integrated photonic devices is continuously developing. Four-wave mixing (FWM), as an important third-order nonlinear optical process, can be used for wavelength conversion and optical signal processing in integrated waveguides, and has received intense investigations on different platform [1]. Materials such as silicon, silicon dioxide, silicon-rich silica, silicon nitride, and nonlinear polymer have been used for FWM applications. The refractive index of silicon is higher, and the optical band gap (1.1 eV) is smaller. However, the conversion efficiency and signal processing speed of the FWM effect in pure silicon are limited by the nonlinear loss (two photon absorption induced free carrier absorption) [2,3]. Silicon dioxide has a smaller refractive index, but the nonlinear index is very low (2.7×10^{-20} m^2/W at 1550 nm), thus it needs a long waveguide to achieve high nonlinear effect. As an alternative choice, silicon nitride waveguides have a moderately high index contrast, a relatively high Kerr nonlinearity, and a negligible two photon absorption around telecom wavelengths, which makes it a potential candidate for photonic integration platforms [4–7]. Moreover, silicon nitride has a wide transparent window ranging from visible to near infrared wavelengths [8]. As a consequence, excellent silicon nitride-based nonlinear optical applications have been demonstrated, including optical frequency combs [9,10], broadband supercontinuum generation [11], self-phase modulation [12], and FWM-based wavelength conversion [13]. Usually, silicon nitride thin films can be prepared by PECVD or low-LPCVD [4].

The dispersion properties of the waveguides are significant for the ultrafast nonlinear applications [14–17]. The lack of dispersion management might lower the wavelength conversion efficiency and limit the tuning range of the FWM process. In principle, near-zero dispersion over a wavelength range is required to achieve phase matching for high FWM conversion efficiencies [18,19]. Note that other third-order nonlinear applications, such as optical frequency combs, broadband supercontinuum generation, self-phase modulation, and cross-phase modulation also require dispersion engineering for phase matching. Both the material dispersion and waveguide dispersion contribute to the total dispersion. The waveguide dispersion can be tailored by controlling the structure parameters of the waveguide [10–12,17]. The phase matching conditions are mainly dominated by second-order dispersion, which is usually described by the dispersion coefficient D.

In this paper, systematic analyses on the dispersion properties of PECVD and LPCVD silicon nitride channel waveguides are conducted, and the geometric parameters of the waveguides to obtain near-zero dispersions for phase matching are optimized. The fabricated silicon nitride micro-ring resonator with optimized parameter shows near-zero dispersion of -0.07 ps/m/nm at 1550 nm and a flat dispersion profile over a 130 nm wavelength range. FWM results show that the optimized silicon nitride waveguide exhibits a high conversion efficiency larger than -12 dB.

2. Waveguide Dispersion Analysis Principle

The silicon nitride channel waveguide is designed on silicon substrate with a SiO_2 claddings. The thicknesses of the bottom and top claddings are 3 μm. The width and the thickness of the core are denoted by w and t, respectively. The relationships between the effective refractive index and the group refractive index and the dispersion coefficient can be expressed by following equations [20]:

$$n_g = n_{\text{eff}} - \lambda \frac{dn_{\text{eff}}}{d\lambda} n_g = n_{\text{eff}} - \lambda \frac{dn_{\text{eff}}}{d\lambda} \quad (1)$$

$$D(\lambda) = \frac{1}{c}\frac{dn_g}{d\lambda} = -\frac{\lambda}{c}\frac{d^2 n_{\text{eff}}}{d\lambda^2} = -\frac{2\pi c}{\lambda^2}\beta_2 \quad (2)$$

where n_{eff} is the effective refractive index, n_g is the group refractive index, λ is the wavelength, c is the speed of the light, β_2 is the group velocity dispersion. The effective refractive index of the waveguide depends on the waveguide width. thickness and the material refractive index. By analyzing the spectrum of n_{eff}, the group refractive index and dispersion of the waveguide can be extracted according to Equations (1) and (2).

3. Results and Discussion

3.1. Waveguide Dispersion Optimiation

The PECVD silicon nitride is deposited at temperature of 350 °C with a gas flow ratio of SiH_4:NH_3:N_2=6:3:189 sccm. The LPCVD silicon nitride is prepared by using dichlorosilane (DCS) and ammonia (NH_3) at 800 °C with a DCS to NH_3 ratio of 0.3 [17]. The material refractive indices are shown in Figure 1. The data will be used in effective refractive index and dispersion calculation. The CVD deposited silicon nitride film can be silicon-rich or nitrogen-rich. Due to the different deposition temperature and gas ratio, there is a relatively large difference between refractive indices of silicon nitride films prepared by PECVD and LPCVD [21,22].

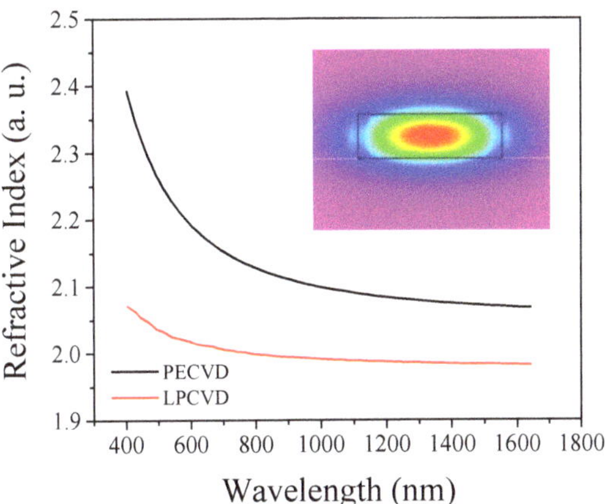

Figure 1. The refractive index of silicon nitride prepared by PECVD and LPCVD. The insert is the optical field distribution of the fundamental TE mode, here, the width, thickness, and refractive index of silicon nitride are 710 nm, 1250 nm and 2.0, respectively.

3.1.1. Dispersion of Silicon Nitride Thin Film Waveguide of PECVD

In this study, only the dispersion of the fundamental TE mode is studied because TM modes become TE modes if the waveguide rotates 90°. The filed distribution of the silicon nitride channel waveguide is calculated by using 3D finite element method (3D-FEM). The insert in Figure 1 shows the field distribution of the fundamental mode of the PECVD silicon nitride waveguide at 1550 nm with a width of 710 nm and a thickness of 1250 nm. It can be seen that the optical power is well confined in the core of the waveguide.

The effective refractive index of the mode is obtained from the mode calculation. Figure 2 shows the effective refractive index and dispersion of the waveguides with different width and thickness at the wavelength of 1550 nm. Note that the dispersion includes the waveguide dispersion and material dispersion. It can be seen that the effective refractive index of the waveguide increases with the width and thickness increasing because more power is confined in the waveguide core while the cross-section is larger.

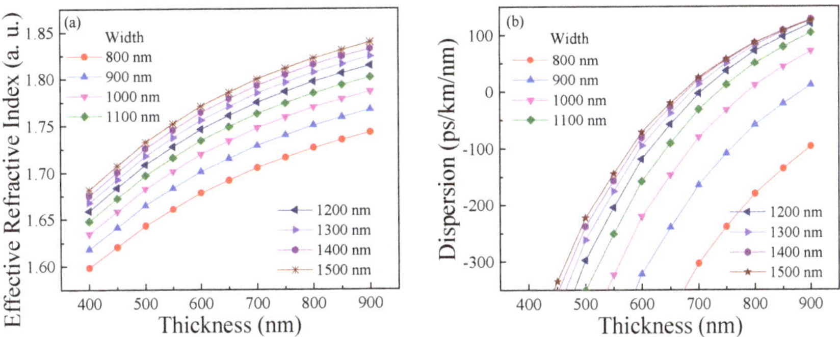

Figure 2. (**a**) The effective refractive index, and (**b**) the dispersion of PECVD silicon nitride waveguide with different widths and thicknesses.

The dispersion increases with the width and thickness increasing. The dispersion increases slowly with the width while the width is larger than 1200 nm. This is because the optical mode distribution is well confined in the waveguide core and weakly affected by the width increase. For the same reason, the thickness dependence of the dispersion will exhibit a similar characteristic. For TE mode, the dispersion changes faster with the thickness than with the width. However, the optimization of the width is easier than that of the thickness because it can be simply realized by designing the photolithograph pattern. Moreover, a very large thickness of the silicon nitride film will add much difficulty to the dry etching of the device. Therefore, a tradeoff between the device fabrication and the dispersion engineering should be made. While the width changing is not enough to get a target dispersion especially an anomalous dispersion ($D > 0$), the thickness should be changed during experimental optimization.

According to Figure 2b, near-zero dispersions can be easily obtained while the thickness is around 700 nm and the width is around 1200 nm. We choose the waveguide parameters as $t = 710$ nm and $w = 1250$ nm which offers an obvious fabrication tolerance. Thus, the dispersion of the selected dimension is slightly larger than zero which might satisfy the phase matching requirement of FWM. It should be noted that the core dimension of the waveguide does not meet the condition of single-mode propagation. However, the power of the high order mode distributes to the edge and attenuates very fast due to transmission loss. Therefore, the high order mode dispersion of the waveguide can be ignored and the dimension supports effective nonlinear interaction between the fundamental mode and the materials. The spectrum of the dispersion of the selected waveguide is shown in Figure 3. The dispersion is near zero (±20 ps/km/nm) over a 100 nm wavelength range around 1550 nm, and is close to zero (~1.16 ps/km/nm) at 1550 nm, which can offer an excellent phase matching condition.

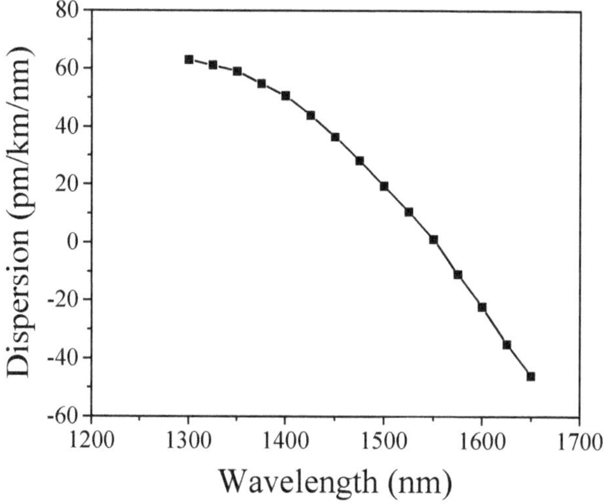

Figure 3. The dispersion as a function of wavelength of PECVD silicon nitride waveguide with optimized thickness of 710 nm and width of 1250 nm to get a near-zero dispersion around the wavelength of 1.55 μm.

3.1.2. Dispersion of Silicon Nitride Thin Film Waveguide of LPCVD

In this section, we analyze the dispersion of channel waveguides of silicon nitride deposited by using LPCVD. Figure 4a shows the dispersion of the waveguide with different width and thickness at the wavelength of 1550 nm. It can be seen that the dispersion increases with the width and thickness. The zero dispersion point can be found on the curve of $w = 1100$ nm. Therefore, the waveguide with thickness of 790 nm and width of 1100 nm was selected, which are close to the optimized parameters of PECVD silicon nitride waveguide. The dispersion spectrum of the optimized waveguide is shown in Figure 4b. It shows flat and low dispersion over a wide wavelength range, and a near-zero dispersion of 0.97 ps/km/nm at the wavelength of 1550 nm is obtained.

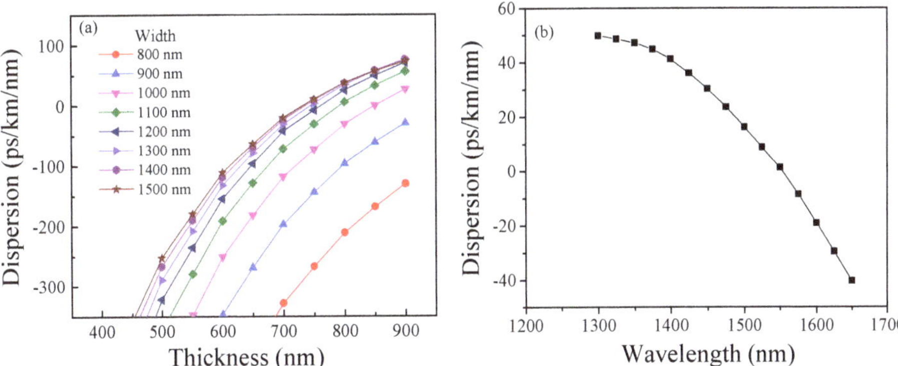

Figure 4. The dispersion of LPCVD silicon nitride waveguide (**a**) with different widths and thicknesses, and (**b**) with an optimized thickness of 790 nm and width of 1100 nm.

Compared Figures 2 and 4, it can be seen that PECVD and LPCVD silicon nitride waveguides exhibit similar dispersion characteristics. Even though the refractive index characteristics are different, near-zero dispersion can be obtained to meet the phase matching for FWM, in which the thickness is in the range of 700 nm to 800 nm and the width is in the range of 1100 nm to 1300 nm. If the dispersion is optimized for TM mode in silicon nitride waveguide, a larger waveguide thickness (>1 μm) will be required because of the rotational relationship between TM and TE modes. The larger thickness might add more difficulty to the film deposition and cause more side wall roughness during etching [4].

3.2. Waveguide Fabrication and Experimental Characterization

The dispersion of waveguides can be experimentally analyzed by measuring the spectrum of an interference waveguide structure such as an MZI interferometer or a micro-ring resonator. Therefore, we fabricate a micro-ring resonator on a standard 4-inch silicon wafer with a 3 μm-thick SiO_2 layer according to the optimized parameters. The radium of the ring, the waveguide width, the gap between the bus waveguide and the micro-ring are 250 μm, 1250 nm and 150 nm, respectively. The 710 nm-thick silicon nitride film is firstly deposited on the wafer at a temperature of 350 °C with a gas flow ratio of SiH_4:NH_3:N_2 = 6:3:189 sccm by using the PECVD (SAMCO, PD-220NL). A 500 nm-thick positive photoresist ARP6200 is then spin-coated on the wafer at the speed of 1500 rpm. After baking at 150 °C for 1 min, the waveguide structure is then patterned by electronic beam lithography (EBL, Elionix, ELS-G100) and etched 710 nm onto the silicon nitride layer by ion coupling plasma (ICP, SAMCO, RIE-400iPB) with the gas of CHF_3. The gap flow, bias power and ICP power are 6 sccm, 50 W, and 50 W, respectively. After removing the photoresist, a 3 μm-thick SiO_2 layer is then deposited on the sample as a top cladding at 150 °C by using liquid source chemical vapor deposition (LSCVD, SAMCO, PD100ST) [23]. Figure 5 illustrates the schematic illustration, micrograph and the SEM image in the coupling zone of the micro-ring resonator. The light coupling was conducted by using waveguide grating couplers which offer excellent design and application flexibilities [24]. The grating couplers are optimized at 1550 nm and the coupling efficiency is about −3.5 dB. The fabricated device agrees well with the design.

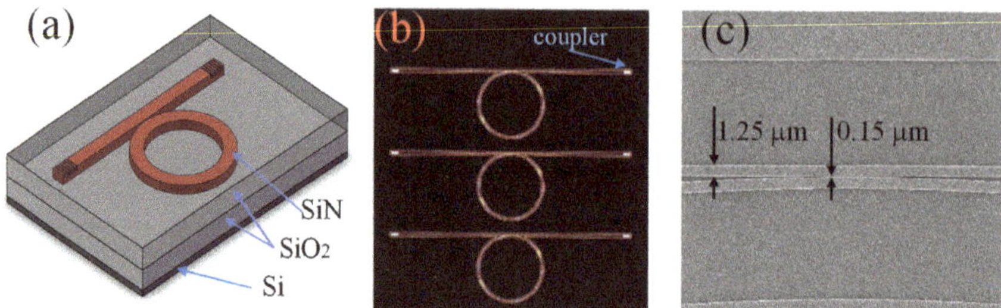

Figure 5. (**a**) The schematic illustration, (**b**) micrograph, and (**c**) SEM image in the coupling region of the micro-ring resonators.

In order to measure the transmission spectrum of the micro-ring resonator, light from a tunable laser (SANTEC TSL550) is pumped into the input waveguide. The wavelength is swept over a range of 130 nm and the spectrum is recorded by an optical power meter (SANTEC MPM200) with 1 pm wavelength resolution and shown in Figure 6a. The insert shows the experimental setup. According to the spectrum data, the free spectral range (*FSR*) is extracted and shown in Figure 6b. It can be seen that *FSR* increases gradually with the wavelength. Note that, the transmission loss of the waveguide is estimated to be 1.82 dB/cm by measuring the fiber-to fiber losses of waveguides with different lengths.

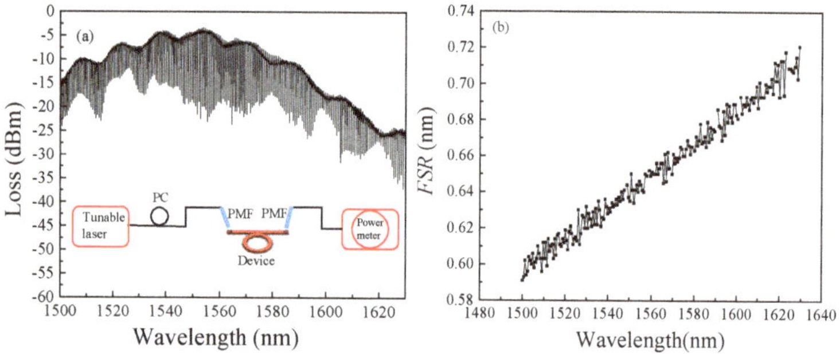

Figure 6. (**a**) The measured transmission spectrum and (**b**) the *FSR* of the silicon nitride micro-ring resonator. The insert is the experimental setup. PC: polarization controller, PMF: polarization maintain fiber.

The relationship between the group refractive index and *FSR* can be written as [25]:

$$n_g = \frac{\lambda^2}{2\pi RFSR} \tag{3}$$

where R is the radius of the ring. According to Equation (3), the group refractive index is calculated and shown in Figure 7a. The red line is the polynomial fitting of the data. By substituting the fitting into Equation (2), the dispersion is calculated and shown in Figure 7b. It shows a flat and low dispersion profile within −0.04 to −0.1 ps/m/nm from wavelength 1500 nm to 1630 nm (a range of 130 nm). Near-zero dispersion of −0.07 ps/m/nm at the wavelength of 1550 nm is achieved. The dispersion decreases with the wavelength which is similar to the simulation results shown in Figures 3 and 4. The wavelength dependence of the dispersion is near linear in Figure 7b, which is a bit different to those shown in Figures 3 and 4 at short wavelengths. It can be explained that the material dispersion which is very linear in the wavelength range as shown in Figure 1 contributes much to the total dispersion of the fabricated device due to the relatively small waveguide size of the micro-ring resonator.

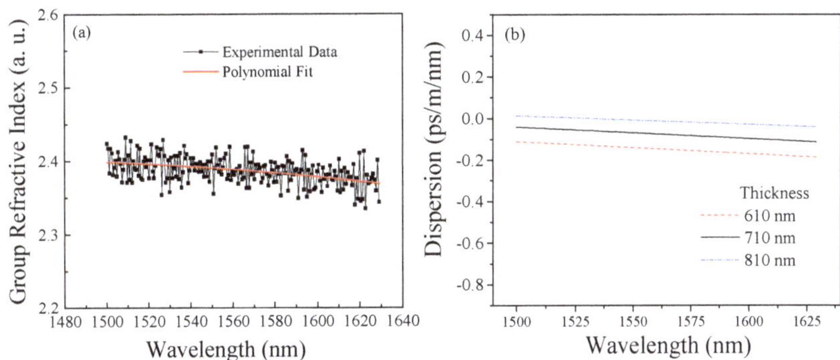

Figure 7. (**a**) The calculated group refractive index, and (**b**) dispersion of the silicon nitride micro-ring resonator.

Dispersions of the waveguides with thicknesses of 610 nm and 810 nm are also measured and shown in Figure 7b. Simulation and experiment results show that near-zero anomalous dispersion ($D > 0$) can be realized by tailoring the dimensions of the waveguides. Sometimes, anomalous dispersion is required for phase matching [10,18]. Especially in the case of large loss existing, the anomalous dispersion can compensate the phase mismatching caused by the linear and nonlinear losses. As demonstrated before, only increasing the width might be incapable of resulting in an anomalous dispersion if the thickness is small. In order to get a near-zero anomalous dispersion, the thickness of 810 nm or larger values is suitable.

3.3. Phase Mismatch and Conversion Efficiency of the Waveguide

The phase mismatch of the waveguide for FWM can be expressed as [26–28]:

$$\kappa = \Delta\beta + 2\gamma P_P = \beta_s + \beta_c - 2\beta_p + 2\gamma P_P \approx \beta_2(\omega_s - \omega_p)^2 + 2\gamma P_P \tag{4}$$

where $\beta_{p,s,c}$ and $\omega_{p,s,c}$ are the wavenumbers and frequencies of the interaction optical waves, P_p is the pump power. the first and second terms, $\Delta\beta$ and $2\gamma P_p$, are referred to as linear and nonlinear phase mismatch, respectively. γ is the waveguide nonlinear coefficient and is defined as [29]:

$$\gamma = \frac{2\pi n_2}{\lambda A_{\mathrm{eff}}} \tag{5}$$

where n_2 is the nonlinear refractive index, A_{eff} is effective mode area of the waveguide. According to the mode distribution, γ is calculated to be 1.19 $\mathrm{m^{-1}W^{-1}}$ at 1550 nm by assuming $n_2 = 2.6 \times 10^{-19}$ $\mathrm{m^2W^{-1}}$ for silicon nitride.

The phase mismatch depends on the dispersion, frequency difference, waveguide nonlinear coefficient and input pump power. Because the value of γ is positive and not too large, a near-zero dispersion especially anomalous dispersion can make the phase mismatch close to zero. Figure 8 shows the phase mismatch of the waveguide with different input pump power. The near-zero phase mismatch (± 0.2 $\mathrm{cm^{-1}}$) is obtained over a 30 nm wavelength range near the pump wave, which is attributed to the optimized near-zero anomalous dispersion of the silicon nitride waveguide. It also shows a weak phase mismatch dependence on the pump power, which means that the conversion efficiency can be effectively improved by increasing the pump power.

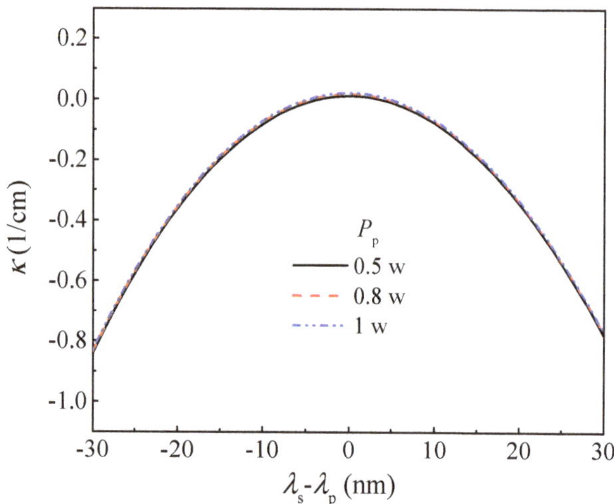

Figure 8. The calculated phase mismatch of the PECVD silicon nitride waveguide at different pump power. Here, the pump wavelength is 1550 nm. The width and height of the waveguide are 1250 nm and 710 nm, respectively.

The coupled equations of the FWM process can be expressed as [27,30]:

$$\frac{dA_p}{dz} = -\frac{1}{2}\alpha A_p + j\gamma|A_p|^2 A_p \quad (6)$$

$$\frac{dA_s}{dz} = -\frac{1}{2}\alpha A_s + 2j\gamma|A_p|^2 A_s + j\gamma A_i^* A_p^2 exp(-j\Delta\beta z) \quad (7)$$

$$\frac{dA_i}{dz} = -\frac{1}{2}\alpha A_i + 2j\gamma|A_p|^2 A_i + j\gamma A_s^* A_p^2 exp(-j\Delta\beta z) \quad (8)$$

where $A_{p,s,c}$ are the amplitude of the interaction optical waves, α is the transmission loss factor, z is the coordinate along the transmission direction. The conversion efficiency (CE) defined as the power ratio of the output idler power to signal power are calculated. Figure 9 shows the CE dependence on the waveguide length with an input pump power of 0.5 W. It can be seen that CE as high as -12 dB can be obtained with a length of 13 cm. The CE increases slowly while the length is larger than 13 cm because the accumulated transmission loss in a very long waveguide attenuates the generated idler power. Moreover, the optical power in a very long waveguide changes obviously during transmission due to the nonnegligible loss. The phase mismatch increases with the transmission distance. Therefore, as can be seen in Figure 9, the CE changes more obviously with the input signal wavelength while the length is larger. The CE decreases with the input signal wavelength decreasing because the wavelength difference between the pump and signal waves enlarges the phase mismatch as shown in Figure 8.

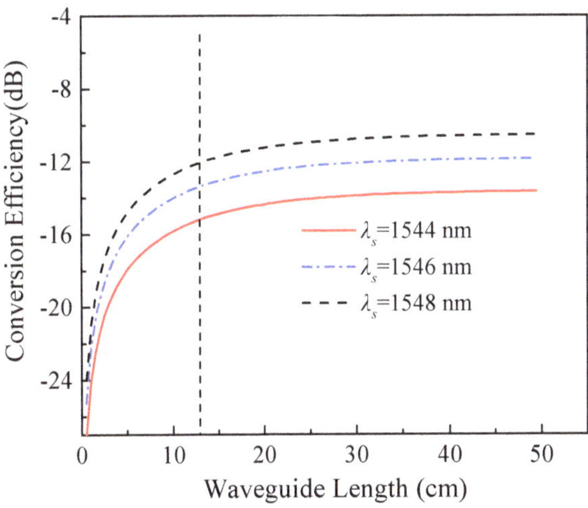

Figure 9. The calculated conversion efficiency of the FWM process of the PECVD silicon nitride waveguide. Here, the pump power and pump wavelength are 0.5 W and 1550 nm, respectively.

4. Conclusions

In this paper, we analyzed the group velocity dispersion of the silicon nitride waveguides. In order to find the near-zero dispersion points for the phase matching of FWM-based application, we optimize the core dimension of the silicon nitride waveguide. The effective refractive index and the dispersion of the waveguide increase with the width and thickness increasing, and the dispersion decreases with the wavelength increasing. The optimum core dimension of the PECVD silicon nitride waveguide is 710 nm $\times$ 1250 nm and the corresponding dispersion is 1.16 ps/km/nm. The optimized core dimension of the LPCVD silicon nitride waveguide is 790 $\times$ 1100 nm and the corresponding dispersion is 0.97 ps/km/nm. Results show that the dimensions of the PECVD and LPCVD silicon nitride for the phase matching of FWM are similar.

The dispersion of a PECVD silicon nitride micro-ring resonator is experimentally characterized with the core dimensions of t = 710 nm and w = 1250 nm. Results show that a near-zero dispersion of -0.07 ps/m/nm at the wavelength of 1550 nm is obtained which is consistent with the theoretical results. Owing to the optimized dispersion of the waveguides, near-zero phase mismatch over a wavelength range of 30 nm is obtained, which ensures high CEs larger than -12 dB with a low input pump power of 0.5 W in a 13 cm long silicon nitride waveguide.

Author Contributions: Conceptualization: J.H. and G.-W.L.; data curation: Y.H. (Yaping Hong) and J.H.; methodology: J.H.; software: Y.H. (Yaping Hong) and Y.H. (Yixiao Hong); validation: Y.H. (Yaping Hong), J.H., and G.-W.L.; writing—original draft: Y.H. (Yaping Hong) and J.H.; writing—review and editing: Y.H. (Yixiao Hong), J.H., and G.-W.L.; funding acquisition: G.-W.L. All authors have read and agreed to the published version of the manuscript.

Funding: This research received no external funding.

Institutional Review Board Statement: Not applicable.

Informed Consent Statement: Not applicable.

Data Availability Statement: Not applicable.

Conflicts of Interest: The authors declare no conflict of interest.

References

1. Helt, L.G.; Liscidini, M.; Sipe, J.E. How Does It Scale? Comparing Quantum and Classical Nonlinear Optical Processes in Integrated Devices. *J. Opt. Soc. Am. B* **2012**, *29*, 2199–2212. [CrossRef]
2. Wu, C.; Huang, J.; Ou, D.; Liao, T.; Chiu, Y.; Shih, M.; Lin, Y.; Chu, A.; Lee, C. Efficient Wavelength Conversion with Low Operation Power in a Ta_2O_5-Based Micro-Ring Resonator. *Opt. Lett.* **2017**, *42*, 4804–4807. [CrossRef]
3. McMillan, J.F.; Yu, M.; Kwong, D.; Wong, C.W. Observation of Four-Wave Mixing in Slow-Light Silicon Photonic Crystal Waveguides. *Opt. Express* **2010**, *18*, 15484–15497. [CrossRef]
4. Yakuhina, A.; Kadochkin, A.; Svetukhin, V.; Gorelov, D.; Generalov, S.; Amelichev, V. Investigation of Side Wall Roughness Effect on Optical Losses in a Multimode Si_3N_4 Waveguide Formed on a Quartz Substrate. *Photonics* **2020**, *7*, 104. [CrossRef]
5. Moss, D.J.; Morandotti, R.; Gaeta, A.L.; Lipson, M. New CMOS-Compatible Platforms Based on Silicon Nitride and Hydex for Nonlinear Optics. *Nat. Photonics* **2013**, *7*, 597–607. [CrossRef]
6. Ikeda, K.; Saperstein, R.E.; Alic, N.; Fainman, Y. Thermal and Kerr Nonlinear Properties of Plasma-Deposited Silicon Nitride/Silicon Dioxide Waveguides. *Opt. Express* **2008**, *16*, 12987–12994. [CrossRef]
7. Boller, K.; van Rees, A.; Fan, Y.; Mak, J.; Lammerink, R.E.M.; Franken, C.A.A.; van der Slot, P.J.M.; Marpaung, D.A.I.; Fallnich, C.; Epping, J.P.; et al. Hybrid Integrated Semiconductor Lasers with Silicon Nitride Feedback Circuits. *Photonics* **2020**, *7*, 4. [CrossRef]
8. Feng, J.; Akimoto, R.A. Three-Dimensional Silicon Nitride Polarizing Beam Splitter. *IEEE Photonics Technol. Lett.* **2014**, *26*, 706–709. [CrossRef]
9. Herr, T.; Hartinger, K.; Riemensberger, J.; Wang, C.Y.; Gavartin, E.; Holzwarth, R.; Gorodetsky, M.L.; Kippenberg, T.J. Universal Formation Dynamics and Noise of Kerr-Frequency Combs in Microresonators. *Nat. Photonics* **2012**, *6*, 480–487. [CrossRef]
10. Pfeiffer, M.H.P.; Herkommer, C.; Liu, J.; Guo, H.; Karpov, M.; Lucas, E.; Zervas, M.; Kippenberg, T.J. Octave-Spanning Dissipative Kerr Soliton Frequency Combs in Si_3N_4 Microresonators. *Optica* **2017**, *4*, 684–691. [CrossRef]
11. Halir, R.; Okawachi, Y.; Levy, J.S.; Foster, M.A.; Lipson, M.; Gaeta, A.L. Ultrabroadband Supercontinuum Generation in a CMOS-Compatible Platform. *Opt. Lett.* **2012**, *37*, 1685–1687. [CrossRef]
12. Tan, D.T.H.; Ikeda, K.; Sun, P.C.; Fainman, Y. Group Velocity Dispersion and Self Phase Modulation in Silicon Nitride Waveguides. *Appl. Phys. Lett.* **2010**, *96*, 611016. [CrossRef]
13. Kruckel, C.J.; Torres-Company, V.; Andrekson, P.A.; Spencer, D.T.; Bauters, J.F.; Heck, M.J.R.; Bowers, J.E. Continuous Wave-Pumped Wavelength Conversion in Low-Loss Silicon Nitride Waveguides. *Opt. Lett.* **2015**, *40*, 875–878. [CrossRef]
14. Yao, Z.; Wan, Y.; Bu, R.; Zheng, Z. Improved Broadband Dispersion Engineering in Coupled Silicon Nitride Waveguides with a Partially Etched Gap. *Appl. Opt.* **2019**, *58*, 8007–8012. [CrossRef] [PubMed]
15. Guo, H.; Herkommer, C.; Billat, A.; Grassani, D.; Zhang, C.; Pfeiffer, M.H.P.; Weng, W.; Bres, C.; Kippenberg, T.J. Mid-infrared Frequency Comb via Coherent Dispersive Wave Generation in Silicon Nitride Nanophotonic Waveguides. *Nat. Photonics* **2018**, *12*, 496. [CrossRef]
16. Okawachi, Y.; Lamont, M.R.E.; Luke, K.; Carvalho, D.O.; Yu, M.; Lipson, M.; Gaeta, A.L. Bandwidth Shaping of Microresonator-Based Frequency Combs via Dispersion Engineering. *Opt. Lett.* **2014**, *39*, 3535–3538. [CrossRef] [PubMed]
17. Kruckel, C.J.; Fulop, A.; Ye, Z.; Andrekson, P.A.; Torres-Company, V. Optical Bandgap Engineering in Nonlinear Silicon Nitride Waveguides. *Opt. Express* **2017**, *25*, 15370–15380. [CrossRef] [PubMed]
18. Liu, Q.; Gao, S.; Li, Z.; Xie, Y.; He, S. Dispersion Engineering of a Silicon-Nanocrystal-Based Slot Waveguide for Broadband Wavelength Conversion. *Appl. Opt.* **2011**, *50*, 1260–1265. [CrossRef]
19. Zhang, L.; Yue, Y.; Xiao-Li, Y.; Wang, J.; Beausoleil, R.G.; Willner, A.E. Flat and Low Dispersion in Highly Nonlinear Slot Waveguides. *Opt. Express* **2010**, *18*, 13187–13193. [CrossRef]
20. Sun, X.; Dai, D.; Thylén, L.; Wosinski, L. Double-Slot Hybrid Plasmonic Ring Resonator Used for Optical Sensors and Modulators. *Photonics* **2015**, *2*, 1116–1130. [CrossRef]
21. Baets, R.; Subramanian, A.Z.; Clemmen, S.; Kuyken, B.; Bienstman, P.; Le Thomas, N.; Roelkens, G.; Van Thourhout, D.; Helin, P.; Severi, S. Silicon Photonics: Silicon Nitride Versus Silicon-On-Insulator. In Proceedings of the 2016 Optical Fiber Communications Conference and Exhibition (OFC), Anaheim, CA, USA, 20–22 March 2016; p. Th3J.1.
22. Mao, S.C.; Tao, S.H.; Xu, Y.L.; Sun, X.W.; Yu, M.B.; Lo, G.Q.; Kwong, D.L. Low Propagation Loss SiN Optical Waveguide Prepared by Optimal Low-Hydrogen Module. *Opt. Express* **2008**, *16*, 20809–20816. [CrossRef] [PubMed]
23. Cheng, X.; Hong, J.; Spring, A.M.; Yokoyama, S. Fabrication of a High-Q Factor Ring Resonator Using LSCVD Deposited Si_3N_4 Film. *Opt. Mater. Express* **2017**, *7*, 2182–2187. [CrossRef]
24. Hong, J.; Spring, A.M.; Qiu, F.; Yokoyama, S. A High Efficiency Silicon Nitride Waveguide Grating Coupler with a Multilayer Bottom Reflector. *Sci. Rep.* **2019**, *9*, 12988. [CrossRef]
25. Yin, Y.; Yin, X.; Zhang, X.; Yan, G.; Wang, Y.; Wu, Y.; An, J.; Wang, L.; Zhang, D. High-Q-Factor Silica-Based Racetrack Microring Resonators. *Photonics* **2021**, *8*, 43. [CrossRef]
26. Hansryd, J.; Andrekson, P.A.; Westlund, M.; Li, J.; Hedekvist, P. Fiber-Based Optical Parametric Amplifiers and Their Applications. *IEEE J. Sel. Top. Quantum Electron.* **2002**, *8*, 506–520. [CrossRef]
27. Gao, S.M.; Li, Z.Q.; Zhang, X.Z. Power-Attenuated Optimization for Four-Wave Mixing-Based Wavelength Conversion in Silicon Nanowire Waveguides. *J. Electromagn. Wave* **2010**, *24*, 1255–1265. [CrossRef]
28. Jia, L.; Geng, M.; Zhang, L.; Yang, L.; Chen, P.; Wang, T.; Liu, Y. Wavelength Conversion Based on Degenerate-Four-Wave-Mixing with Continuous-Wave Pumping in Silicon Nanowire Waveguide. *Opt. Commun.* **2009**, *282*, 1659–1663. [CrossRef]

29. An, L.; Liu, H.; Sun, Q.; Huang, N.; Wang, Z. Wavelength Conversion in Highly Nonlinear Silicon-Organic Hybrid Slot Waveguides. *Appl. Opt.* **2014**, *53*, 4886–4893. [CrossRef] [PubMed]
30. Jin, B.; Yuan, J.; Yu, C.; Sang, X.; Wei, S.; Zhang, X.; Wu, Q.; Farrell, G. Efficient and Broadband Parametric Wavelength Conversion in a Vertically Etched Silicon Grating without Dispersion Engineering. *Opt. Express* **2014**, *22*, 6257–6268. [CrossRef] [PubMed]

Article

An Optical Analog-to-Digital Converter with Enhanced ENOB Based on MMI-Based Phase-Shift Quantization

Yue Liu , Jifang Qiu *, Chang Liu, Yan He , Ran Tao and Jian Wu

School of Electronic Engineering, Beijing University of Posts and Telecommunications, Beijing 100876, China; LYmoon@bupt.edu.cn (Y.L.); liuchang345@bupt.edu.cn (C.L.); heyan@bupt.edu.cn (Y.H.); taoran2018@bupt.edu.cn (R.T.); jianwu@bupt.edu.cn (J.W.)

* Correspondence: jifangqiu@bupt.edu.cn

Abstract: An optical analog-to-digital converter (OADC) scheme with enhanced bit resolution by using a multimode interference (MMI) coupler as optical quantization is proposed. The mathematical simulation model was established to verify the feasibility and to investigate the robustness of the scheme. Simulation results show that 20 quantization levels (corresponding to 4.32 of effective number of bits (ENOB)) are realized by using only 6 channels, which indicates that the scheme requires much fewer quantization channels or modulators to realize the same amount of ENOB. The scheme is robust and potential for integration.

Keywords: analog-to-digital converter; optical phase-shift quantization; ENOB

Citation: Liu, Y.; Qiu, J.; Liu, C.; He, Y.; Tao, R.; Wu, J. An Optical Analog-to-Digital Converter with Enhanced ENOB Based on MMI-Based Phase-Shift Quantization. *Photonics* **2021**, *8*, 52. https://doi.org/10.3390/photonics8020052

Received: 25 January 2021
Accepted: 11 February 2021
Published: 14 February 2021

Publisher's Note: MDPI stays neutral with regard to jurisdictional claims in published maps and institutional affiliations.

1. Introduction

High-speed analog-to-digital converters (ADC) play an important role in modern signal processing, such as high-frequency radar systems [1–5], wireless communications [6,7], and image processing [8,9], etc. However, with more and more wideband applications, such as wireless and radar communications especially when the bandwidth of radar signals is over 10 GHz, traditional electronic ADCs have been unable to meet the requirements of high speed due to timing jitter, thermal noise and uncertainty of electronic components, and so on. In recent years, with the rapid development of photonic technology, using photonic technology to break through the bottleneck of electronic ADC has become a research hotspot in the field of optoelectronic technology [10]. The advantage of using photonic technology in ADC is that the sampling rate can be up to 100GS/s (samples per second) or even higher. Moreover, the timing jitter of the optical pulse can reach the level of femtosecond [11,12], which is two orders of magnitude smaller than that of the electrical sampling pulse. In addition, the pulse width can be compressed to the order of picoseconds or even femtoseconds by using pulse compression technology [13]. Therefore, optical analog-to-digital conversion (OADC) technology has significant advantages compared with traditional electronic ADC technology. In general, OADC has great potential not only in terms of high speed, high precision, and wide bandwidth, etc., but also in commercial applications due to the great development of photonic integration. Thus, it has become one of the main development directions of high-speed ADC.

In last decades, lots of optical-sampling electrical-quantization [14,15] and optical-sampling optical-quantization schemes have been reported, which can be divided into two categories, i.e. intensity-to-wavelength optical quantization and phase-shift optical quantization (PSOQ). Intensity-to-wavelength optical quantization [16,17] usually realizes intensity to wavelength conversion by using nonlinear effects via nonlinear devices, like highly nonlinear fiber (HNLF). However, the length of HNLF is usually up to km level and the optical power required is relatively large to excite the nonlinear effect. On the other hand, the PSOQ scheme, which was firstly proposed by Taylor [18] and then developed in [19–23], typically uses electro-optic modulators with multiplied half-wave voltages. The advantages of Taylor's and its developed schemes are potentiality of on-chip integration and have low requirement of input optical power due to operation principle based on optical interference. In general, in Taylor's and its developed schemes, optical quantization of N bit requires N modulators. To further increase the quantization bits, the half-wave voltage of the last modulator must be significantly reduced. Therefore, we previously proposed a cascade step-size (CSS) multimode interference (MMI) coupler-based phase quantization scheme [24] requiring only one phase modulator (PM). In the scheme reported in [24], in order to realize N bit, only one phase modulator and a CSS-MMI with 2^{N-1} output channels are needed. This scheme avoids the increasing requirement of the number of modulators when N increases in the schemes developed from Taylor's scheme, and eliminates the need for ultra-low half-wave voltage of modulators. However, when N is larger, the number of output channels (K) increases significantly (for example, when N is 4.17 bit, K is 9; whereas N is 5.09 bit, K is 17, and so on), resulting in enormous difficulties in design and fabrication of CSS-MMI.

In this paper, in order to obtain effective number of bits (ENOB) as high as possible in OADC with the required number of modulators and optical output channels as fewer as possible, we propose a new PSOQ scheme of OADC, which effectively increases the number of quantization bits but requires much fewer optical channels and modulators. We numerically demonstrate the scheme with 6 optical channels and 2 modulators by using a 1×4 MMI, a 3×5 MMI, a PM, and a Mach-Zehnder modulator (MZM), which can achieve 4.32 bit. Compared with our previously proposed scheme [24], which needs 9 optical channels to achieve 4.17 bit, this new scheme is easier for design and fabrication. Moreover, we investigated the performance of the proposed scheme by building up a simulation system, and found that compared with the previous scheme, the proposed scheme not only has much fewer output channels, but also has more robustness in presence of timing jitter, intensity noise of optical sampling pulses, and MMI power imbalance.

2. Principle of Operation

The schematic diagram of the proposed OADC is shown in Figure 1. It consists of an optical sampling module, an optical quantization module, and an electrical back-end processing module. The optical sampling module is composed of a pulse laser, a 1×4MMI (MMI^{1st}), and a PM. The optical quantization module is composed of a 3×5 MMI (MMI^{2nd}) and an MZM. The digital signal processing module is composed of photo-detector (PD) arrays and electric comparator arrays.

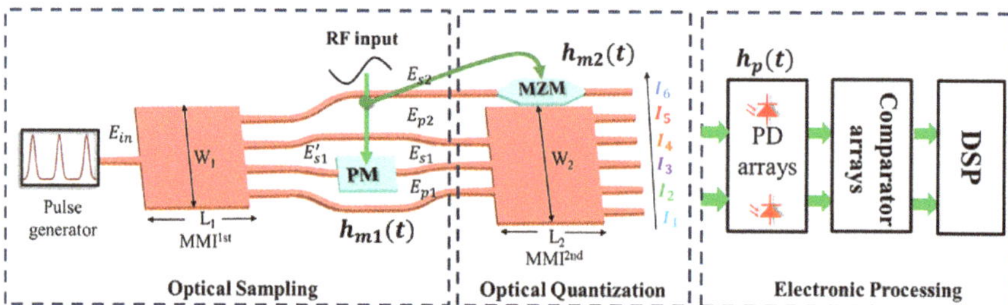

Figure 1. Schematic diagram of the proposed 4.32-bit optical analog-to-digital converter (OADC). RF input: Radio-frequency input. PM: phase modulator, MZM: Mach-Zehnder modulator, PD: photo-detector, DSP: digital signal processing.

Firstly, in the optical sampling module, assuming that the injected high-speed pulses emitted by the pulse laser are solitary pulses with Gauss shape, can be represented as

$$E_{in} = \sum_{n=-\infty}^{n=+\infty} R_G(t - nT_S) \tag{1}$$

where $R_G(t) = \sqrt{P_{max}} \cdot e^{-2\cdot(\frac{t}{T_{FWHM}}^{2m})} \cdot e^{(2\pi f_0 t + \varphi_0)}$, T_S ($T_S = 1/f_S$) is the sampling rate. P_{max} is the peak power of the pulse. The full width at half maximum power (FWHM) pulse duration is T_{FWHM} and m denotes the order of the super-Gaussian function. The carrier frequency of the optical pulse is defined by emission frequency f_0 and the initial phase is φ_0. Therefore, the schematic of waveforms of the sampling pulses is shown in Figure 2a.

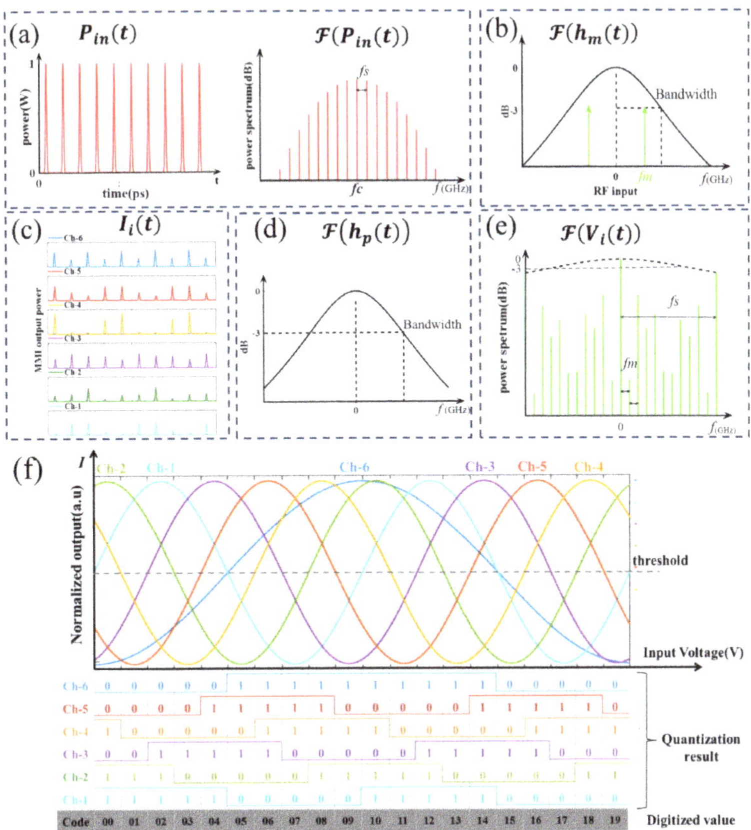

Figure 2. The proposed OADC with sinusoidal RF input. (**a**) Waveforms and frequency spectrum the high-speed optical sampling pulses. (**b**) Spectral response of both modulators. (**c**) The outputs of 6 channels under single-frequency sinusoidal input signal. (**d**) Spectral response of PDs. (**e**) power spectrum. (**f**) Transmission curves of 6 optical channels and digital output codes with input voltage.

The high-speed pulses E_{in} are divided into four channels with equal power by MMI^{1st}, the outputs of MMI^{1st} can be described as

$$\begin{bmatrix} E_{s2} \\ E_{p2} \\ E'_{s1} \\ E_{p1} \end{bmatrix} = \frac{1}{2} \begin{bmatrix} e^{j0} \\ e^{j\frac{\pi}{2}} \\ e^{j\frac{\pi}{2}} \\ e^{j0} \end{bmatrix} \cdot E_{in} \tag{2}$$

After splitting, optical pulse E'_{s1} passes through PM which is driven by input signal $V_{in}(t)$, and becomes E_{s1} as

$$E_{s1} = E'_{s1} \cdot e^{j\Delta\Phi_{s1}(t)},\ \Delta\Phi_{s1}(t) = h_{m1}(t) * (V_{in}(t) + V_{b1}) \tag{3}$$

where $h_{m1}(t)$ is the equivalent model of a phase modulator which reflects the high frequency characteristics of the phase modulator. The electronic frequency response is shown as Figure 2b, where f_m is the frequency of the RF (Radio-frequency) input. $\Delta\Phi_{s1}(t)$ can be equivalent to $\frac{\pi(V_{in}(t)+V_{b1})}{V_\pi}$, where V_π and V_{b1} are the half-wave voltage and bias voltage of PM, respectively. $\Delta\Phi_{s1}(t)$ is induced by $V_{in}(t)$. Then E_{s1} is injected at the middle position of MMI^{2nd}, while E_{p1} and E_{p2} are injected at positions as

$$x = \frac{K \pm 2}{2K} W_2 \tag{4}$$

where K is the number of MMI^{2nd} output channels, here $K = 5$. After carefully choosing the dimensions of MMI^{2nd}, K images of equal intensities can be formed at MMI^{2nd} output. The output power can be expressed as

$$I_i(t) = \begin{cases} \frac{2\sqrt{2}P}{K}\left(\frac{3\sqrt{2}}{4} + cos\left(\Delta\Phi_{s1}(t) + \frac{i}{2}\cdot\frac{2\pi}{K} - \frac{3\pi}{4}\right)\right) & for\ even\ i \\ \frac{2\sqrt{2}P}{K}\left(\frac{3\sqrt{2}}{4} + cos\left(\Delta\Phi_{s1}(t) - \frac{i-1}{2}\cdot\frac{2\pi}{K} - \frac{3\pi}{4}\right)\right) & for\ odd\ i \end{cases} \tag{5}$$

where $i = 1 \ldots 5$. From Equation (5), we can see that the output power of each channel varies sinusoidally with $V_{in}(t)$ as shown in Figure 2f. This is similar to [24] and acts as quantizer which can realize 10 quantization levels (corresponding to 3.32 bit). The output pulses of each channel are shown in Figure 2c. Based on the principle in [24], in order to extend the ENOB, the 3 × 5 MMI should be replaced by a 3 × 9 MMI, thus OADC can be extended to 18 quantization levels, which means a 4.17 bit OADC with 9 output channels is obtained.

In this paper, we propose a new scheme to extend the ENOB by introducing a MZM. As shown in Figure 1, the MZM with a half-wave voltage of $2V_\pi$ is introduced parallel to the PM, with E_{s2} as input. The output power of MZM can be expressed as

$$I_6(t) = \frac{P}{2}(1 + cos(\Delta\Phi_{s2}(t))), \Delta\Phi_{s2}(t) = h_{m2}(t) * (V_{in}(t) + V_{b1}) \tag{6}$$

where $h_{m2}(t)$ is the numerical model of MZM and reflects the high frequency characteristics of the MZM. $\Delta\Phi_{s2}(t)$ is equivalent to $\frac{\pi(V_{in}(t)+V_{b2})}{V_\pi}$, V_{b2} is the bias voltage applied to MZM.

After the optical-to-electrical (O/E) conversion performed by the PD arrays, we can obtain the electrical signal as

$$V_i(t) = [(I_i(t) + I_n(t))\cdot h_p(t)]\cdot R_L \quad i = 1 \ldots 6 \tag{7}$$

$I_n(t)$ is the noise current caused by PD in each channel, $h_p(t)$ is the equivalent circuit model of PD as shown in Figure 2d and the power spectrum of $V_i(t)$ is Figure 2e. R_L is load resistance that converts current into voltage output.

Next, the detected signals of the 6 output channels are injected into the electric comparator arrays, judged to "0" or "1" by appropriate threshold as

$$D_i(n)|_{t=nT_s} = \begin{cases} V_{oh}, & V_i(t) - V_{th} > V_{in.min} \\ V_{ol}, & V_{th} - V_i(t) > V_{in.min} \quad i = 1 \ldots 6 \\ D_i(n-1), & |V_{th} - V_i(t)| < V_{in.min} \end{cases} \tag{8}$$

Generally, $V_{oh} = 1$ and $V_{ol} = 0$. V_{th} is the threshold voltage of electric comparator arrays, $V_{in.min} = \frac{V_{oh}-V_{ol}}{A_v}$ is the minimum identification precision voltage, which is related to the conversion precision A_v of the comparator. The higher the A_v is, the smaller the $V_{in.min}$ will be, which shows that the comparator has more accurate comparison results.

At last, after processing these digital signals according to the electronic decision rules as shown in Figure 2f, we can get 20 different quantization levels. We can see from Figure 2f that, the transmission periods of Channel 1 to Channel 5 are $2V_{\pi}$. For Channel 1 to Channel 5 we can obtain 10 quantization levels range from 0 to $2V_{\pi}$, and the encoding is repeated within $2V_{\pi}$ to $4V_{\pi}$. Channel 6 is synchronized with other channels, but the transmission period is $4V_{\pi}$. By combining all 6 channels, 20 quantization levels are obtained in the input voltage ranging from 0 to $4V_{\pi}$. Compared with our previously proposed scheme (where 18 quantization levels are obtained with 9 output channels), the proposed scheme needs fewer channels (6 channels) to obtain equivalent quantization levels (i.e., ENOB).

3. Simulation Modelling and Performance Analysis

Based on the above models, the OADC is built up to verify the feasibility and investigate the performance of the proposed OADC scheme, as shown in Figure 3. We analyze the OADC performance based on IEEE standard [25], where ENOB is derived as [25]

$$ENOB = \frac{SINAD - 1.76\,\text{dB}}{6.02} \tag{9}$$

SINAD (signal-to-noise ratio) well reflects the overall dynamic performance of the ADC because it includes all factors that constitute noises and distortions. Based on IEEE standard, SINAD is calculated from the frequency spectrum of the digital output codes of the ADC as

$$SINAD = 10lg\frac{\sigma_x^2}{\sigma_q^2 + \sigma_n^2} \tag{10}$$

where σ_x^2 is the electrical signal power, σ_q^2 and σ_n^2 represent inherent quantization errors and errors caused by other noise, respectively.

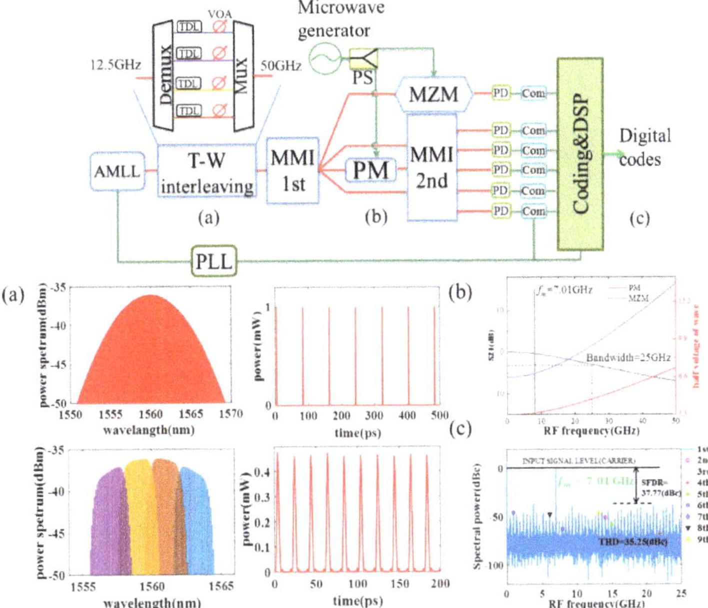

Figure 3. Schematic of and simulation results of the proposed OADC. AMLL: active mode-locked laser; Demux: wavelength division multiplexer. Mux: wavelength division multiplexer. TDL: time delay line; VOA: tunable optical amplifier; PS: power splitter; MMI1st: 1 × 4 MMI. MMI2nd: 3 × 5 MMI. PD: photo-detector. Com: electronic comparators. PLL: phase-locked loop. (**a**) 50 GHz optical pulses are generated by time-wavelength interleaving. (**b**) S21 response and half-wave voltage change with input RF signal. (**c**) FFT spectrum of digital output codes.

We use time-wavelength interleaving (T-W interleaving) to acquire optical sampling pulses with f_s = 50 GHz. The waveforms of the output of active mode-locked laser (AMLL) is shown in Figure 3a, whereas the time-wavelength interleaved sampling pulses are shown in Figure 3a. A sine wave signal of 7.01 GHz was used to act as an electrical analog signal, which is divided into two by power splitter (PS), one of which is used to drive the PM, while the other one is used to drive the MZM. Based on the principle analysis, the ideal ENOB is 4.32 bit. Then we took imperfections in the actual system into account to calculate the ENOB. Table 1 shows the main simulation parameters of the OADC. The parameters chosen in Table 1 are based on the datasheet of commercially available devices.

Table 1. Main parameters and values of the OADC in simulation.

Component		Parameter	Value
Pulse laser		P_{max}/Intensity jitter Sampling rate/Timing jitter T_{FWHM}	0 dBm/2% 12.5 GHz/100 fs 0.5 ps
T-W interleaving	Demux/mux	Bandwidth Channel interleaving	150 GHz 250 GHz
	TDL	Time delay	0 ps 20 ps 40 ps 60 ps
	VOA	Amplification power	Related with each channel
Modulator	PM	Analog bandwidth half-wave voltage	25 GHz 3.3 V(@1 GHz)
	MZM	Analog bandwidth half-wave voltage	25 GHz 6.6 V(@1 GHz)
MMI		Normalized channel power imbalance	0.943 1.010 1.027 0.884 1.136
PDs		Bandwidth R_L Dark current thermal noise shot noise	40 GHz 50 Ω 5 nA Related with bandwidth and resistance Related with pulse power
Comparators		Offset voltage A_v	1 mV 30 dB
PLL		RMS(J_{per})	63 fs

For example, the parameters of AMLL was chosen based on the datasheet of Calmar PS-10-TT series, like timing jitter is 100 fs, the standard deviation of random intensity noise (RIN) is 2% relative to the pulse peak power, and the T_{FWHM} of the Gaussian pulse train is 0.5 ps, peak power is 0 dBm, and wavelength is 1560 nm.

As for modulators, with the increase of input RF frequency, RF and light wave speed mismatch and transmission loss caused by the structural parameters of lithium niobate modulator increase as well, which in turn affects the output optical signal. Therefore, it is important to analyze the performance of OADC under this influence. The RF response of modulator can be represented by S21 response, which decreases with the input frequency increasing. This will affect the OADC output phase $\Delta\Phi_{s1}(t)$ and $\Delta\Phi_{s2}(t)$ as shown in Figure 3b. In our simulation, the half-wave voltages of PM and MZM were set as 3.3 V@1 GHz and 6.6 V@1 GHz referring to the commercial modulator, and the S21 data of commercial devices were taken into account. MMI transmission curves were obtained according to measurement results based on our fabricated MMI [26], which has normalized channel power imbalance of [0.943 1.010 1.027 0.884 1.136], leading to non-uniform quantization and thus decreasing the ENOB of the OADC.

In the electronic processing module, the parameters of PD were chosen according to the silicon germanium PIN structure photodetector mentioned in [27–29], as bandwidth of 40 GHz, a load resistance of 50 Ω, dark current of 5 nA, electronic measurement bandwidth and resistance-related thermal noise, and shot noise related to input optical power. Similarly, the parameters for electrical comparators were set with reference to the offset voltage and minimum conversion accuracy A_v of commercial electrical comparators (ADI, MP582). The electrical components are synchronized with the active mode-locked laser (AMLL) through a phase-locked loop (PLL). The root mean square period jitter (RMS(J_{per})) caused by phase noise of PLL internal electronic circuits is set to 63 fs according to commercial PLL (Texas Instruments, LMX2820).

At last, according to the output digital code and f_s, the frequency of the input signal is calculated according to the FFT calculation result and Equation (9) as shown in Figure 3c, which is consistent with the given input signal f_m, verifying the correctness of the simulation modelling, the SINAD is calculated to be 26.03 dB. Therefore, the ENOB is 4.03 bit, which is a little slighter than the ideal ENOB of 4.32 bit and demonstrates the accuracy of the simulation modelling of the proposed OADC scheme.

Table 2 compares the performance of different reported schemes based on PSOQ, "-" in Table 2 means that CW (continuous-wave) light was used or non-mentioned in the schemes. We can see from Table 2 that our proposed scheme has a simpler structure due to the fact that the optical quantizer is a passive device and has fewer optical channels and accepted number of modulators, which is a potential for on-chip integration (similar optical quantizer with 5 channels has already been experimentally reported [26]).

Table 2. The performance comparisons of OADC schemes.

Schemes	Year	Type of Result	Sampling Rate	Signal Input	Quantization Type	Number of Modulators	Number of Optical Channels	ENOB
[19]	2011	Experimental	-	10 GHz	PSOQ	1	8	3.2
[20]	2011	Simulated	20 GHz	-	PSOQ	1	8	4
[21]	2014	Simulated	-	25 GHz	PSOQ	1	24	5.28
[22]	2018	Experimental	-	400 MHz	PSOQ	3	8	3.31
[24]	2018	Simulated	-	-	PSOQ	1	5	3.27
[23]	2020	Experimental	-	500 MHz	PSOQ	3	10	3.75
This work	2021	Simulated	50 GHz	7.1 GHz	PSOQ	2	6	4.03

Next, in order to investigate the robustness of the OADC, we investigated the factors that RIN and timing jitter of the pulse source, the MMI channel imbalance and signal noise, phase noise of PLL on ENOB.

3.1. RIN and Timing Jitter

The actual optical sampling pulse has timing jitter, RIN, and limited pulse width. These characteristics of the optical pulse cause errors in sampling process, reduce the SINAD of the OADC, and degrade the ENOB. It is important to investigate the effects of RIN and timing jitter, as both time-wavelength interweaving and back-end electrical processing also lead to mismatches and amplitude fluctuations throughout the OADC. Therefore, we investigated the impact of RIN and timing jitter on ENOB. RIN is defined as $\frac{\delta I}{P_{max}}$, where δI is intensity noise, and P_{max} is pulse peak power. Full scale modulation is assumed in the calculation and the amplitude of the RF input is A. Therefore, the error caused by RIN can be substituted into Equation (10) to calculate SINAD

$$\mathrm{SINAD} = \begin{cases} \dfrac{A^2/2}{\frac{\left(\frac{2A}{N}\right)^2}{12} + (2K+1)\left(\frac{\arcsin\frac{\mathrm{RIN}}{1+\mathrm{RIN}}}{2\pi/N}\cdot\frac{2A}{N}\right)^2\cdot\frac{2}{N}} & K=5, N=20 \\ \dfrac{A^2/2}{\frac{\left(\frac{2A}{N}\right)^2}{12} + K\left(\frac{\arcsin\frac{\mathrm{RIN}}{1+\mathrm{RIN}}}{2\pi/N}\cdot\frac{2A}{N}\right)^2\cdot\frac{2}{N}} & K=9, N=18 \end{cases} \quad (11)$$

Timing jitter can be defined as δT. The judgment time of the sampling pulse is $t = \mathrm{n}T_s$. According to the power shape is given by $P(t) = |R_G(t)|^2$, Equation (7), we can calculate the SINAD caused by timing jitter as following

$$\mathrm{SINAD} = \begin{cases} \dfrac{A^2/2}{\frac{\left(\frac{2A}{N}\right)^2}{12} + \sum_{i=1}^{i=K+1} \frac{1}{T_s} \int_0^{T_s} [V_i(t) \cdot |\, P(t-\delta T) - 1|]^2 dt} & K = 5, N = 20 \\ \dfrac{A^2/2}{\frac{\left(\frac{2A}{N}\right)^2}{12} + \sum_{i=1}^{i=K} \frac{1}{T_s} \int_0^{T_s} [V_i(t) \cdot |\, P(t-\delta T) - 1|]^2 dt} & K = 9, N = 18 \end{cases} \quad (12)$$

It can be seen from Equation (12) that SINAD is related to T_s and the frequency f_m of the input signal. $f_s = 100$ GHz is set in the calculation. SINAD degrades when RIN increases, which can be theoretically obtained as shown in Figure 4a. As shown in Figure 4b, SINAD drops faster when f_m increases, because there is a greater risk of error in judgment time.

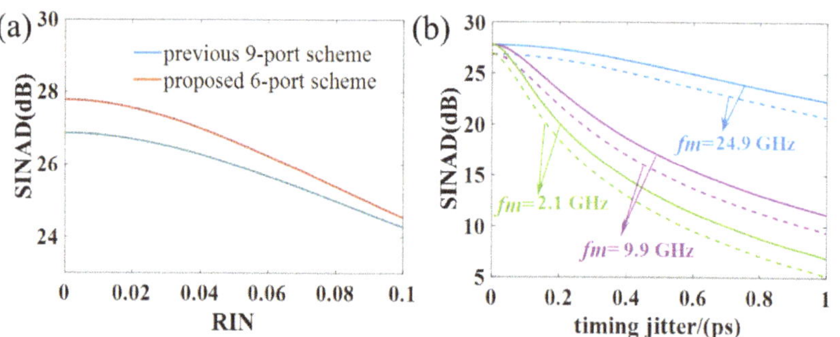

Figure 4. SINAD degradation induced by the high-speed pulse RIN and time jitter. (**a**) Theoretical calculation of the effect of RIN on SINAD. (**b**) Theoretical calculation of the effect of timing jitter on SINAD with different f_m. The straight lines represent the proposed 6-port scheme and the dashed lines represent the previous 9-port scheme.

We demonstrate how RIN and timing jitter affect system performance from the mathematical model. We carried out the simulation operation when sample rate of optical pulse is 100 GHz and $f_m = 24.9$ GHz in our simulation modelling. The final 2^{15} digital codes were selected to calculate FFT and the final ENOB was obtained. Here, P_{max} equals to -5 dBm, we changed RIN from 0 to 10%, timing jitter from 0 ps to 1 ps. The results of ENOB vs. timing jitter and RIN are shown in Figure 5a as a 3D graph. The same investigation was also carried out for our previously 9-port scheme with equivalent ENOB for comparison. The results of the previous 9-port scheme are shown in Figure 5b. We can see from Figure 5 that ENOB decreases greatly with the increase of RIN and timing jitter. When ENOB degrades by 1 bit, in Figure 5a, RIN tolerance is 9.6%, while in Figure 5b RIN tolerance of our previous 9-port scheme is 5.3%. The results indicate that the new proposed OADC has lower requirements on RIN of the sampling pulses. As for timing jitter, we can see that when ENOB degrades by 1 bit, the tolerance of timing jitter is 0.5 ps, while the previous 9-port scheme is 0.4 ps as shown in Figure 5b. The results indicate that the proposed scheme is more robust to timing jitter.

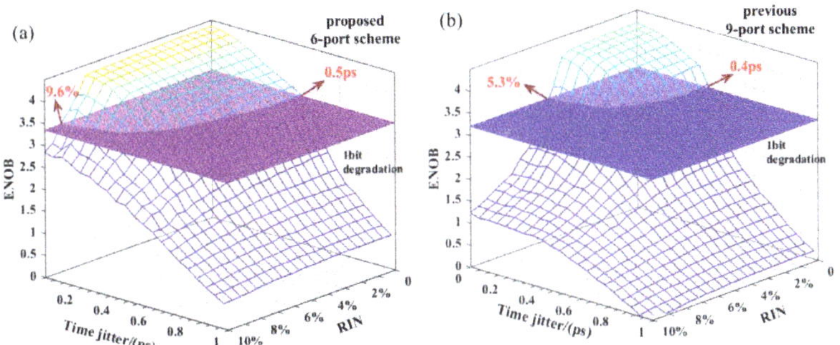

Figure 5. ENOB degradation induced by RIN and time jitter of the optical sampling pulse. (**a**) Simulation results of proposed 6-port scheme. (**b**) Simulation results of previous 9-port scheme.

3.2. Channel Imbalance

Channel equalization refers to the consistency in amplitude of the modulation curve outputs of the quantizer (MMI^{2nd}). The channel imbalance of the MMI^{2nd} may be caused by manufacturing process errors, or design errors. Therefore, it is important to investigate the effect of channel imbalance of MMI^{2nd} on ENOB.

The power imbalance of each channel is expressed by $\delta P_i = \frac{P_i}{(\sum_{i=1}^{K} P_i)/K}$, $i = 1 \ldots K$. Therefore, the SINAD under the error caused by channel imbalance is given by

$$\text{SINAD} = \begin{cases} \dfrac{A^2/2}{\frac{\left(\frac{2A}{N}\right)^2}{12} + 2\sum_{i=1}^{i=K}\left(\frac{\arcsin\frac{\delta P_i}{1+\delta P_i}}{2\pi/N}\cdot\frac{2A}{N}\right)^2\cdot\frac{2}{N}} & K=5, N=20 \\ \dfrac{A^2/2}{\frac{\left(\frac{2A}{N}\right)^2}{12} + \sum_{i=1}^{i=K}\left(\frac{\arcsin\frac{\delta P_i}{1+\delta P_i}}{2\pi/N}\cdot\frac{2A}{N}\right)^2\cdot\frac{2}{N}} & K=9, N=18 \end{cases} \tag{13}$$

where $i = 1 \ldots K$. Since the δP_i of each channel is different, when calculating the effect of δP_i on OADC performance, it can be represented by

$$\text{IM} = \sqrt{\frac{\sum_{i=1}^{K}(\delta P_i - 1)^2}{K}} \tag{14}$$

It can be seen from Equation (13) that SINAD is related to the number of channels of MMI^{2nd}. As shown in Figure 6a, SINAD drops faster with more channels, because there is a greater probability of errors in the judgement. Different IM values are set in the simulation to observe the impact on ENOB of the OADC. In simulation, ENOB was calculated by 50 times for the same IM values. The red curve in Figure 6b shows ENOB degradation induced by IM, the results for previous 9-port scheme are shown as a black curve. We can see that both ENOB degrades with IM increasing, this is because due to the imbalance, the actual quantization step size is different from the ideal quantization step size, which would lead to non-uniform quantization errors and make harmonic noise power increase, resulting in a decrease in ENOB. Furthermore, as shown in Figure 6b, it has been verified by 50-time simulations that if the ENOB degrades by 1 bit, the tolerance of IM of the proposed scheme can be 16%, but the previous 9-port scheme has only 10.9% tolerance. In comparison, the proposed scheme is more robust to the power imbalance of the quantizer. We can also see that under the same IM, because the proposed scheme has fewer ports, the probability of error in the output digital code is smaller than that of the previous 9-port scheme.

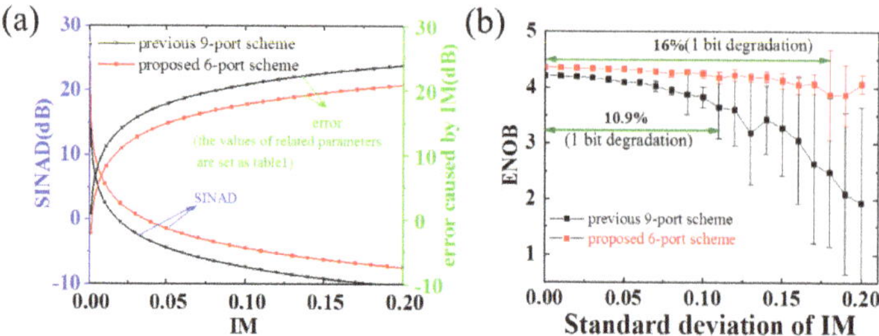

Figure 6. The effect of channel imbalance on system performance. (**a**) Theoretical calculation of the effect of channel imbalance on SINAD. (**b**) Simulation results at the condition of MMI with channel imbalance.

3.3. Electronic Devices Noise

The noise introduced by the electrical devices in the OADC also affects the performance of the OADC. Here we analyze two typical effects on the ENOB of the OADC. One is the noise of the input analog RF signal, the other is the phase noise of the PLL used to synchronize the sampling pulses with comparators and coding circuits, etc.

3.3.1. The Noise of Input Signal

The signal source and RF amplifier may introduce noise. Therefore, we analyzed the noisy signal on the performance of OADC. The signal to noise ratio (SNR) of the analogue signal is expressed as SNR_{in}. Assuming that the noise added to the analogue signal is a Gaussian random signal with a mean value of 0. Therefore, the SINAD under the error caused by noise is given by

$$\text{SINAD} = \begin{cases} \dfrac{A^2/2}{\frac{\left(\frac{2A}{N}\right)^2}{12} + 2(\text{K}+2)\cdot P_{noise}\cdot\frac{2}{N}} & K = 5, N = 20 \\ \dfrac{A^2/2}{\frac{\left(\frac{2A}{N}\right)^2}{12} + K\cdot P_{noise}\cdot\frac{2}{N}} & K = 9, N = 18 \end{cases} \quad (15)$$

where P_{noise} is the power of the noisy signal. It can be seen from Equation (15) that SINAD is related to the quantization level N. We carried out the simulation operation with other parameters unchanged. The results of ENOB vs. P_{noise} are shown in Figure 7a. Dozens of simulations have been carried out to eliminate the contingency of the results, which indicate that the higher the number of quantization levels, the faster the ENOB decreases. That is to say that the noise of the input signal has a greater impact on OADC with bigger ENOB, thus the SNR of the input signal cannot be ignored.

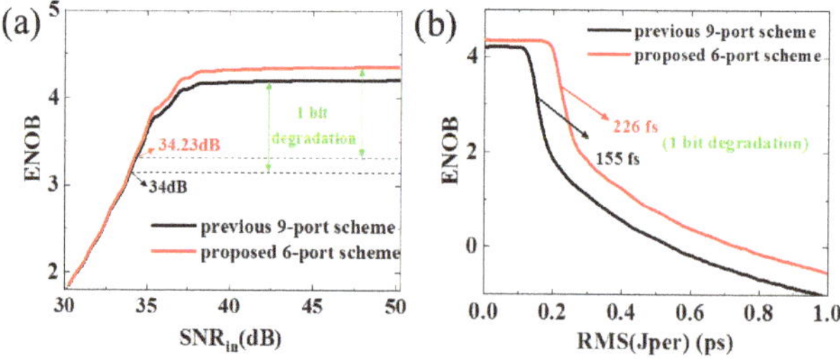

Figure 7. The effect of noisy signal and phase noise of PLL on system performance. (**a**) Simulation results of ENOB under different SNR of input analog signal. (**b**) Simulation results of ENOB under different RMS(J_{per}) of PLL.

3.3.2. Phase Noise of PLL

The reference clock signal of the PLL is the synchronous output of the pulse laser, thus PLL could provide a locked clock to the electronic comparators and coding circuits. However, the clock signal generated from an actual PLL has phase noise ($\theta(t)$), which is equivalent to the jitter of the clock signal (J_{per}), that represents clock non-synchronization. The clock jitter J_{per}, which is a random disturbance signal and described by root mean square value (RMS) as

$$\text{RMS}(J_{per})(seconds) = \frac{1}{2\pi f_c}\sqrt{\theta^2(t)} = \frac{1}{2\pi f_c}\sqrt{2\int_0^{\infty} S_\theta(f)df} \tag{16}$$

where f_c is clock frequency, $S_\theta(f)$ is the power spectrum of phase noise.

Different RMSs of clock jitter are added in the theoretical calculation and simulation to analyze the influence on OADC. According to $P(t)$, Equation (7), we can calculate the SINAD caused by clock jitter J_{per} as following

$$\text{SINAD} = \begin{cases} \dfrac{A^2/2}{\frac{\left(\frac{2A}{N}\right)^2}{12} + \sum_{i=1}^{i=K+1} \frac{1}{T_s}\int_0^{T_s} [|V_i(t-J_{per})\cdot P(t-J_{per}) - V_i(t)\cdot P(t)|]^2 dt} & K=5, N=20 \\ \dfrac{A^2/2}{\frac{\left(\frac{2A}{N}\right)^2}{12} + \sum_{i=1}^{i=K} \frac{1}{T_s}\int_0^{T_s} [|V_i(t-J_{per})\cdot P(t-J_{per}) - V_i(t)\cdot P(t)|]^2 dt} & K=9, N=18 \end{cases} \tag{17}$$

It can be seen from Equation (17) that SINAD drops faster when J_{per} increases, because there is a greater risk of error in judgment time. We changed RMS(J_{per}) from 0 to 1 ps, and maintained the other parameters. The results of ENOB vs. clock jitter are shown in Figure 7b. When ENOB degrades by 1 bit, RMS(J_{per}) tolerance is 226 fs for the proposed scheme, while it is 155 fs for our previous 9-port scheme. The results indicate that the new proposed OADC has lower requirements on phase noise of PLL due to fewer channels than previous 9-port scheme.

4. Conclusions

In this paper, a new OADC scheme with ENOB enhanced by introducing an MZM to increase the optical quantization levels has been proposed and demonstrated in simulation. Based on the operation principle, 20 quantization levels (corresponding to 4.32 of ENOB) are realized by using only 6 channels, which indicates that the scheme requires much fewer quantization channels or modulators to realize the same amount of ENOB. A numerical analysis and proof-of-concept demonstrations were carried out to verify the feasibility and robustness of the scheme. Simulation results show that 4.03 of ENOB is realized for 7.01 GHz RF signal under 50 GHz sampling rate, which indicates that the proposed scheme requires much fewer quantization output channels or modulators to realize the same amount ENOB, compared with our previous 9-port scheme. We analyzed the influence of RIN and timing jitter of sampling pulses and the channel imbalance of MMI^{2nd} on ENOB of both the proposed OADC scheme and the previous 9-port scheme. The simulation results show that the proposed scheme can tolerate up to 9.6% of RIN and up to 0.5 ps of timing jitter of sampling pulses, as well as 16% power imbalance of the MMI output channels when ENOB degrades by 1bit; in contrast, the previous 9-port scheme can tolerate 5.3% of RIN and 0.4 ps of timing jitter of sampling pulses, and 10.9% power imbalance of the MMI. Based on the operation principle, if the proposed scheme extends to a 5 bit operation, only 10 ports are needs to achieve 36 quantization levels (ideally 5.17 bit); in contrast, 17 ports are needed in our previous 9-port scheme to achieve 34 quantization levels (ideally 5.08 bit). When the ENOB continues to increase, the advantage of fewer ports requirement in the proposed scheme becomes even more significant. This advantage makes our scheme more compact and more robustness, and therefore easier for integration and more practical for system applications.

In future, to realize an OADC chip based on our proposed scheme, there exist two main challenges, one of which is heterogeneous integration of optical devices, the other challenge is optoelectronic integration. However, as we know, the above two challenges, optical heterogeneous integration and optoelectronic integration, have been some of the most interesting and promising research fields throughout the world for decades [30,31]. With the breakthrough of the two integration technologies, our OADC would be practically realized.

Author Contributions: Conceptualization, Y.L., J.Q. and C.L.; investigation, Y.L., R.T.; data curation, Y.L. and J.Q.; writing—original draft preparation, Y.L., J.Q.; writing—review and editing, Y.L., J.Q. and Y.H.; visualization, Y.L., J.Q., C.L., J.W.; supervision, Y.L., J.Q., Y.H. and J.W.; project administration, Y.L., J.Q. and J.W. All authors have read and agreed to the published version of the manuscript.

Funding: This work is partly supported by NSFC program (61875020, 61935003), National Key Research and Development Program of China (2019YFB1803601).

Institutional Review Board Statement: Not applicable.

Informed Consent Statement: Not applicable.

Data Availability Statement: The data presented in this study are available on request from the corresponding author. The data are not publicly available due to the data also forms part of an ongoing study.

Acknowledgments: The authors express their appreciation to the anonymous reviewers for their valuable suggestions.

Conflicts of Interest: The authors declare no conflict of interest.

References

1. Wang, S.; Wu, G.; Su, F.; Chen, J. Simultaneous Microwave Photonic Analog-to-Digital Conversion and Digital Filtering. *IEEE Photonics Technol. Lett.* **2018**, *30*, 343–346. [CrossRef]
2. Yang, G.; Zou, W.; Yuan, Y.; Chen, J. Wideband signal detection based on high-speed photonic analog-to-digital converter. *Chin. Opt. Lett.* **2018**, *16*. [CrossRef]
3. Ma, Y.; Liang, D.; Peng, D.; Zhang, Z.; Zhang, Y.; Zhang, S.; Liu, Y. Broadband high-resolution microwave frequency measurement based on low-speed photonic analog-to-digital converters. *Opt. Express* **2017**, *25*, 2355–2368. [CrossRef] [PubMed]
4. Ghelfi, P.; Ma, L.; Wu, X.; Yao, M.; Willner, A.E.; Bogoni, A. All-Optical Parallelization for High Sampling Rate Photonic ADC in Fully Digital Radar Systems. In Proceedings of the Optical Fiber Communication Conference, San Diego, CA, USA, 21 March 2010; pp. 1–3.
5. Ghelfi, P.; Laghezza, F.; Scotti, F.; Serafino, G.; Capria, A.; Pinna, S.; Onori, D.; Porzi, C.; Scaffardi, M.; Malacarne, A.; et al. A fully photonics-based coherent radar system. *Nature* **2014**, *507*, 341–345. [CrossRef] [PubMed]
6. Lee, J.; Weiner, J.; Chen, Y. A 20-GS/s 5-b SiGe ADC for 40-Gb/s Coherent Optical Links. *IEEE Trans. Circuits Syst. I Regul. Pap.* **2010**, *57*, 2665–2674. [CrossRef]
7. Varzaghani, A.; Kasapi, A.; Loizos, D.N.; Paik, S.; Verma, S.; Zogopoulos, S.; Sidiropoulos, S. A 10.3-GS/s, 6-Bit Flash ADC for 10G Ethernet Applications. *IEEE J. Solid-State Circuits* **2013**, *48*, 3038–3048. [CrossRef]
8. Yoshioka, K.; Kubota, H.; Fukushima, T.; Kondo, S.; Ta, T.T.; Okuni, H.; Watanabe, K.; Hirono, M.; Ojima, Y.; Kimura, K.; et al. A 20-ch TDC/ADC Hybrid Architecture LiDAR SoC for 240 $\times$ 96 Pixel 200-m Range Imaging With Smart Accumulation Technique and Residue Quantizing SAR ADC. *IEEE J. Solid-State Circuits* **2018**, *53*, 3026–3038. [CrossRef]
9. Jin, X.; Liu, Z.; Yang, J. New Flash ADC Scheme With Maximal 13 Bit Variable Resolution and Reduced Clipped Noise for High-Performance Imaging Sensor. *IEEE Sens. J.* **2013**, *13*, 167–171. [CrossRef]
10. Khilo, A.; Spector, S.J.; Grein, M.E.; Nejadmalayeri, A.H.; Holzwarth, C.W.; Sander, M.Y.; Dahlem, M.S.; Peng, M.Y.; Geis, M.W.; DiLello, N.A.; et al. Photonic ADC: Overcoming the bottleneck of electronic jitter. *Opt. Express* **2012**, *20*, 4454–4469. [CrossRef]
11. Nejadmalayeri, A.H.; Grein, M.; Khilo, A.; Wang, J.P.; Sander, M.Y.; Peng, M.; Sorace, C.M.; Ippen, E.P.; Kärtner, F.X. A 16-fs aperture-jitter photonic ADC: 7.0 ENOB at 40 GHz. In Proceedings of the CLEO:2011—Laser Applications to Photonic Applications, Baltimore, Maryland, 1 May 2011; p. CThI6.
12. Benedick, A.J.; Fujimoto, J.G.; Kärtner, F.X. Optical flywheels with attosecond jitter. *Nat. Photonics* **2012**, *6*, 97–100. [CrossRef]
13. Esman, D.J.; Wiberg, A.O.J.; Alic, N.; Radic, S. Highly Linear Broadband Photonic-Assisted Q-Band ADC. *J. Lightwave Technol.* **2015**, *33*, 2256–2262. [CrossRef]
14. Gevorgyan, H.; Al Qubaisi, K.; Dahlem, M.S.; Khilo, A. Silicon photonic time-wavelength pulse interleaver for photonic analog-to-digital converters. *Opt. Express* **2016**, *24*, 13489–13499. [CrossRef]
15. Zhang, H.; Zou, W.; Yang, G.; Chen, J. Dual-output modulation in time-wavelength interleaved photonic analog-to-digital converter based on actively mode-locked laser. *Chin. Opt. Lett.* **2016**, *14*. [CrossRef]

16. Nagashima, T.; Hasegawa, M.; Konishi, T. 40 GSample/s All-Optical Analog to Digital Conversion With Resolution Degradation Prevention. *IEEE Photonics Technol. Lett.* **2017**, *29*, 74–77. [CrossRef]
17. Satoh, T.; Takahashi, K.; Matsui, H.; Itoh, K.; Konishi, T. 10-GS/s 5-bit Real-Time Optical Quantization for Photonic Analog-to-Digital Conversion. *IEEE Photonics Technol. Lett.* **2012**, *24*, 830–832. [CrossRef]
18. Taylor, H.F. An electrooptic analog-to-digital converter. *Proc. IEEE* **1975**, *63*, 1524–1525. [CrossRef]
19. Chi, H.; Li, Z.; Zhang, X.; Zheng, S.; Jin, X.; Yao, J.P. Proposal for photonic quantization with differential encoding using a phase modulator and delay-line interferometers. *Opt. Lett.* **2011**, *36*, 1629–1631. [CrossRef] [PubMed]
20. Chen, Y.; Chi, H.; Zheng, S.; Zhang, X.; Jin, X. Differentially Encoded Photonic Analog-to-Digital Conversion Based on Phase Modulation and Interferometric Demodulation. *IEEE Photonics Technol. Lett.* **2011**, *23*, 1890–1892. [CrossRef]
21. Kang, Z.; Zhang, X.; Yuan, J.; Sang, X.; Wu, Q.; Farrell, G.; Yu, C. Resolution-enhanced all-optical analog-to-digital converter employing cascade optical quantization operation. *Opt. Express* **2014**, *22*, 21441–21453. [CrossRef]
22. He, H.; Chi, H.; Yu, X.; Jin, T.; Zheng, S.; Jin, X.; Zhang, X. An improved photonic analog-to-digital conversion scheme using Mach–Zehnder modulators with identical half-wave voltages. *Opt. Commun.* **2018**, *425*, 157–160. [CrossRef]
23. Yang, S.; Liu, Z.; Chi, H.; Zeng, R.; Yang, B. A Photonic Digitization Scheme With Enhanced Bit Resolution Based on Hierarchical Quantization. *IEEE Access* **2020**, *8*, 150242–150247. [CrossRef]
24. Tian, Y.; Qiu, J.; Huang, Z.; Qiao, Y.; Dong, Z.; Wu, J. On-chip integratable all-optical quantizer using cascaded step-size MMI. *Opt. Express* **2018**, *26*, 2453–2461. [CrossRef]
25. IEEE. *IEEE Standard for Terminology and Test Methods for Analog-to-Digital Converters*; IEEE Std 1241–2010 (Revision of IEEE Std 1241–2000); IEEE: New York, NY, USA, 2011; pp. 1–139. [CrossRef]
26. Liu, C.; Qiu, J.; Tian, Y.; Tao, R.; Liu, Y.; He, Y.; Zhang, B.; Li, Y.; Wu, J. Experimental demonstration of an optical quantizer with ENOB of 3.31 bit by using a cascaded step-size MMI. *Opt. Express* **2021**, *29*, 2555–2563. [CrossRef]
27. Virot, L.; Benedikovic, D.; Szelag, B.; Alonso-Ramos, C.; Karakus, B.; Hartmann, J.-M.; Le Roux, X.; Crozat, P.; Cassan, E.; Marris-Morini, D.; et al. Integrated waveguide PIN photodiodes exploiting lateral Si/Ge/Si heterojunction. *Opt. Express* **2017**, *25*, 19487–19496. [CrossRef] [PubMed]
28. Chen, H.; Galili, M.; Verheyen, P.; Heyn, P.D.; Lepage, G.; Coster, J.D.; Balakrishnan, S.; Absil, P.; Oxenlowe, L.; Campenhout, J.V.; et al. 100-Gbps RZ Data Reception in 67-GHz Si-Contacted Germanium Waveguide p-i-n Photodetectors. *J. Lightwave Technol.* **2017**, *35*, 722–726. [CrossRef]
29. Assefa, S.; Xia, F.; Vlasov, Y.A. Reinventing germanium avalanche photodetector for nanophotonic on-chip optical interconnects. *Nature* **2010**, *464*, 80–84. [CrossRef]
30. Bowers, J.E. Heterogeneous Photonic Integration on Silicon. In Proceedings of the 2018 European Conference on Optical Communication (ECOC), Rome, Italy, 23–27 September 2018; pp. 1–3.
31. Poulton, C.V.; Byrd, M.J.; Raval, M.; Su, Z.; Li, N.; Timurdogan, E.; Coolbaugh, D.; Vermeulen, D.; Watts, M.R. Large-scale silicon nitride nanophotonic phased arrays at infrared and visible wavelengths. *Opt. Lett.* **2017**, *42*, 21–24. [CrossRef] [PubMed]

Communication

Programmable High-Resolution Spectral Processor in C-band Enabled by Low-Cost Compact Light Paths

Zichen Liu [1], Chao Li [2], Jin Tao [3] and Shaohua Yu [3,*]

1 Wuhan National Laboratory for Optoelectronics, Huazhong University of Science and Technology and Wuhan Research Institute of Post and Telecommunication, Wuhan 430074, China; zchliu@wri.com.cn
2 Information Materials and Intelligent Sensing Laboratory of Anhui Province, Institutes of Physical Science and Information Technology, Anhui University, Hefei 230601, China; chao.li@ahu.edu.cn or seven_chao0620@163.com
3 National Information Optoelectronics Innovation Center, Wuhan Research Institute of Posts and Telecommunications, Wuhan 430074, China; taojin@wri.com.cn
* Correspondence: shaohua.yu@cict.com

Received: 2 October 2020; Accepted: 3 December 2020; Published: 7 December 2020

Abstract: The flexible photonics spectral processor (PSP) is an indispensable element for elastic optical transmission networks that adopt wavelength division multiplexing (WDM) technology. The resolution and system cost are two vital metrics when developing a PSP. In this paper, a high-resolution 1 × 6 programmable PSP is investigated and experimentally demonstrated by using low-cost compact spatial light paths, which is enabled by a 2 K (1080p) liquid crystal on silicon (LCoS) and two cascaded transmission gratings with a 1000 line/mm resolution. For each wavelength channel, the filtering bandwidth and power attenuation can be manipulated independently. The total insertion loss (IL) for six ports is in the range of 5.9~9.4 dB over the full C-band. The achieved 3-dB bandwidths are able to adjust from 6.2 GHz to 5 THz. Furthermore, multiple system experiments utilizing the proposed PSP, such as flexible spectral shaping and optical frequency comb generation, are carried out to validate the feasibility for the WDM systems.

Keywords: programmable; photonics spectral processor (PSP); liquid crystal on silicon (LCoS); high-resolution and low-cost; optical frequency combs (OFC)

1. Introduction

With the exponential growth of communication traffic driven by high-definition TVs, online social networks, cloud computing and the Internet of things (IoT), there is an unprecedented requirement on high-capacity data transmission over optical transport networks [1]. The next generation wavelength division multiplexing (WDM) optical networks are required to be elastic. They must be capable of dynamically and efficiently allocating a flexible amount of optical spectra with different modulation baud rates to separated outputs [2]. To maximize the network efficiency with limited fiber bandwidths, a new G.694.1 standard has been established by the International Telecommunication Union. The new provision has a 12.5 GHz bandwidth slot width, which helps to reduce the optical protective band in order to improve the transmission efficiency in the previous G.692 specification. Moreover, advanced modulation techniques such as multi- carrier orthogonal frequency division multiplexing (OFDM) and single-carrier Nyquist shaping combined with high-order quadrature amplitude modulation (QAM) formats are expected. This scheme is expected to result in promising solutions for realizing highly-efficient future super-channel optical networking where multiple carriers are very close to each other with minimal guard-bands [3,4]. Both programmable high-resolution photonics spectral processors (PSPs) and wavelength selective switches (WSSs) can switch optical signals flexibly. Benefiting from the adoption of PSP and WSS, flexible-grid reconfigurable optical add/drop multiplexers

(ROADM) realize the functionalities of the supporting programmable channel bandwidth and provide a fine wavelength tuning of sub-GHz, i.e., 6.25 GHz [5].

PSP provides a higher resolution than WSS, excluding the fact that they achieve the same functions of filtering and optical power manipulation. To achieve cost efficiency and a satisfactory performance, both WSSs and PSPs need to consider the factors such as cost, volume, and insertion loss (IL). It has been proposed that one replaces the expensive M × N WSS with multiple 1 × K WSSs for the partially directional high-degree ROADMs [6]. Generally, researchers hope to increase the number of ports and the degree of integration by sharing an expensive liquid crystal on silicon (LCoS) device and bulk optical components [7–9]. It is believed that designing a high-resolution 1×K PSP with lower cost and compact size can be helpful in decreasing the total cost of an M × N PSP, since it is compatible with the multiple integration solution demonstrated in [8]. Due to its multiple functionalities, the programmable PSP can be implemented in future ultra-high capacity optical systems and networks.

Generating the shape of the arbitrary spectrum: The generation of a variable bandwidth and arbitrary spectrum shape is important in the implementation of the system, such as channel shaping, channel selection and channel gain equalization. For example, by programming the shape of the filter into PSP, one can study the cascading effect of the filter on the transmission quality.

Generating proper optical frequency combs: The generation of controllable optical combs is the key point in optical WDM transmission systems. It requires the PSP to have the high-resolution capability of attenuating the optical power arbitrarily and creating special spectral lines even if they have independent shapes with modulation.

Currently, the spatial light modulator (SLM) utilizing LCoS (LCoS-SLM) is the most competing technique adopted in the design of multiple port WSSs [10–14]. LCoS device combing with gratings can be further developed as programmable PSP for processing optical spectrum resources [15]. Recently, demonstrations have been reported on generating a reconfigurable PSP [15–18]. In [15], Gao et al. proposed and demonstrated a flexible PSP based on a 4K LCoS with a minimum pixel spacing of 3.74 μm. However, they just developed a one-input one-output filter with an achieved minimum 3-dB bandwidth of 10 GHz. Moreover, the price of the employed high-resolution LCoS-SLM is quite expensive. In 2019, their group also demonstrated a reconfigurable PSP, that provided a minimum bandwidth of 12.5 GHz in the C-band by using digital micro-mirror devices. The major drawback is that the achieved tuning resolutions of the center wavelength and IL of this filter are large, being 4.125 GHz (0.033 nm) and 10 dB across the whole C-band, respectively [16]. Rudnick et al. proposed the use of a waveguide grating router (WGR) to implement a high-resolution PSP with a 0.8 GHz optical resolution [17]. However, the achieved free spectral range is just 200 GHz. Besides, to obtain such a high-resolution, large delays within the WGR are required while still maintaining phase accuracy at the output, which is beyond standard fabrication tolerances. Employing spatial light paths, WSS with a high input and output ports has been demonstrated in [8,19,20]. In order to better meet the future development of high-resolution PSP, the designed spatial light paths are required to be compact and have a low IL.

In this paper, we propose low-IL and compact spatial light paths in order to develop a high-resolution low-cost programmable PSP. The concrete scheme is realized by using a 2K LCoS- SLM (The price of 2K LCoS-SLM is less than half of 4K LCoS-SLM) and two cascaded diffraction gratings with a 1000 line/mm resolution. We experimentally demonstrate a 1 × 6 PSP operating across the entire C-band. The achieved minimum 3-dB bandwidth is 6.2 GHz and the largest coverage can be extended to the full C-band of 5 THz spectra by a tuning step of 3.1 GHz. The best and worst ILs for all six ports are less than 6 dB and 9.5 dB, respectively, with about a 1-dB wavelength-dependent loss (WDL). Moreover, arbitrary spectral amplitude manipulations with more than a 20-dB suppression ratio have been experimentally obtained. Compared with the optical signal generator based on commercial products, our proposed method has a better performance for the minimum bandwidth setting and spectral suppression ratio for optical frequency comb (OFC) generation. The achieved results reveal that it is feasible to develop such a programmable and high-resolution PSP by employing commercial cheap devices and compact spatial light paths and to further implement them in a cost-efficient M × N PSP. This is because the employed optical components and the designed compact light paths help to improve the wavelength separation ability, while this does not affect the design of the number of input and output ports of the WSS.

2. Materials and Methods

2.1. Principle of Operation

When designing an optical filter, the center wavelength tuning resolution R_{res} and minimum spectral bandwidth B_{min} are the two key specifications that need to be take into consideration. In this paper, by using two cascaded gratings, the proposed method and light path contribute to achieving a wide dispersed spectrum in space, finally cover larger working length on the LCoS chip. Based on [15], the larger spot size across the y-axis, the smaller the spectral bandwidth. Figure 1 illustrates the operation principle of the LCoS-based programmable PSP. The near parallel input beam is injected onto the transmission gratings. The employed gratings (PING-Sample-083) with a 1000 line/mm resolution feature a high efficiency and high tolerance to the illumination angle of incidence. With a 90% diffraction efficiency, the incidence angle range can be tuned from 46° to 54° [21].

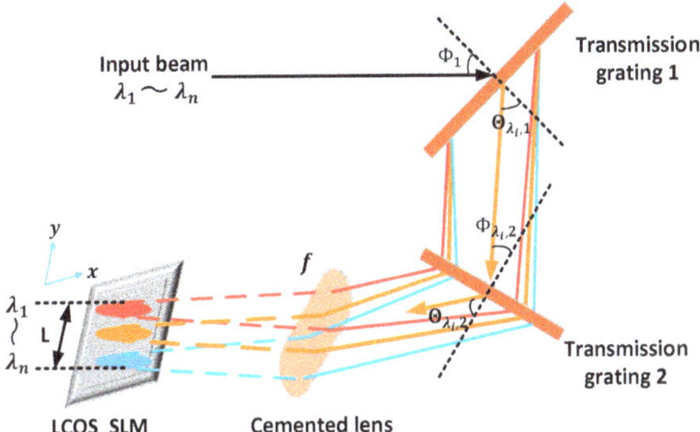

Figure 1. Operation principle of the liquid-on-crystal (LCoS) based programmable PSP using dual-grating.

In the experiment, the incidence angle to the first grating is set at the optimal value of 49.9°, which means that Φ_1 equals to 49.9°. When the input optical beam passes through the transmission grating 1, the wideband signals ($\lambda_1 = 1530$ nm~$\lambda_n = 1570$ nm in the experiment) are spread into angularly separated wavelengths. The first-order diffraction angle of the transmission grating 1 follows the grating equation as follows [22]

$$\sin\Theta_{\lambda_i,1} + \sin\Phi_1 = \frac{\lambda_i}{d} \quad (1)$$

where $\Theta_{\lambda_i,1}$ represents the diffraction angle of the input optical beam with the wavelength of λ_i transmitted after the first grating. d is the groove width of the gratings, which is 1 μm. According to Equation (1), the diffraction angle difference for the wideband signal can be calculated by $\Delta\Theta_1 = \Theta_{\lambda_n,1} - \Theta_{\lambda_1,1} = 53.6° - 49.9° = 3.7°$. Then the angular separated lights are transformed into the second transmission grating. The optimal incidence angle of 49.9° is assigned to the central wavelength of 1550 nm. Thus, the incidence angles for 1530 nm and 1570 nm are set as $49.9° + 3.7°/2 = 51.75°$ and $49.9° - 3.7°/2 = 48.05°$, respectively. The calculated diffraction angle difference after traveling through the second grating is $\Delta\Theta_2 = \Theta_{\lambda_n,2} - \Theta_{\lambda_1,2} = 55.7° - 48.1° = 7.6°$. Therefore, compared with the single grating, the dispersed angle in space is doubled. The transmission grating 2 and LCoS chip are arranged at the front and back focal planes of the cemented lens, whose focal length is $f = 100$ mm. The position distribution L along the y-axis on the LCoS chip is further expressed as

$$L = 2f \cdot \tan(\Delta\Theta_2/2) \tag{2}$$

Thus, based on Equation (2), the theoretical spot size can be calculated as 13.28 mm, which is consistent with the measured result of 12.6 mm. This error of about 5% is acceptable caused by the optical path design and package. The main reason for the specific error is that in the process of packaging and debugging, in order to reduce the fine adjustment of the combined lens of the WDL, the imaging of the spot from the collimator on the LCoS-SLM is changed after passing through the optical subsystem. The input light spots from the collimator array are expanded by a 20-fold cylindrical beam expansion system, resulting in an enlarged elliptical image on the LCoS-SLM. The image is distributed by wavelength in the horizontal direction of the SLM (y-axis).

The employed LCoS chip (Holoeye Pluto) includes an array of 1920 × 1080 pixels with a pitch of 8 μm and effective length of 15.36 mm and can be addressed pixel-by-pixel. The reciprocal linear dispersion along the wavelength axis (y-axis) on the LCoS chip plane is 0.37 GHz/μm. That is to say, one pixel corresponds to a 3-GHz bandwidth. Thus, the minimum resolution of the central channel wavelength R_{res_min} is 3 GHz. If a 4K LCoS-SLM is employed, the minimum resolution can be further enhanced. Theoretically, the spectral bandwidth is an integer multiple of the minimum resolution, in proportion to the number of used pixels in the y-axis. One can note that when the number of cascaded gratings increases to $m > 5$, the diffraction angle difference $\Delta\Theta_m = \Theta_{\lambda_n,m} - \Theta_{\lambda_1,m} > m \cdot \Delta\Theta_1$ and the minimum resolution of the MHz scale can be theoretically achieved.

2.2. *Implementation*

In this section, the implementation and characterization of the 1 × 6 programmable PSP is described.

2.2.1. Optical Path Design Using Zemax

In this paper, we first theoretically study the design of the compact light path based on the Gauss beam coupling equation [23]. The positions and parameters of the used optical elements are calculated. We further optimize them by using the optical path optimization and tolerancing software of Zemax as illustrated in Figure 2. This operation ensures the good coupling efficiency of the Gaussian beams from the input port couples into the multiple output ports and achieves lower values of IL. We then fix and solidify all the optical elements on an optical substrate according to the compact design in the experiment. No mechanical adjustments are needed. It is worth noting that in the optical design the beam spot size expanded by the combination of the cylindrical lenses (Components 2 and 3) matches the sizes of the two gratings, so that both gratings can be fully utilized. Meanwhile, no additional light leakage loss is introduced. The x-arises in Figure 3a,b show the normalized pupil, whose function is to represent the normalized field of view, ranging from −1 to +1. As shown in Figure 3a at the wavelength direction (y-axis), the optical wavelengths of 1530 nm, 1550 nm, and 1570 nm are uniformly distributed on the LCoS plane and the simulated spot size from 1530 nm to 1570 nm is about 12.89 mm, which confirms the measured result of 12.6 mm. At the switching direction (x-axis), the spherical aberration of the optical system is optimized, and no chromatic aberration is caused for the entire C-band wavelengths, since the ray fan curves for 1530 nm, 1550 nm and 1570 nm completely coincide, as shown in Figure 3b. Furthermore, the beam spots formed by different wavelengths are well separated, as shown in Figure 3a, and the respective ellipticity of the beam spot is quite large at the switching direction, which ensures that full use is made of the LCoS pixels (approximately 400 pixels). The elliptical spots on the LCoS plane make it easier to achieve a linear optimization and high control accuracy for the signal attenuation of each wavelength (the minimum power attenuation of 0.1 dB can be achieved).

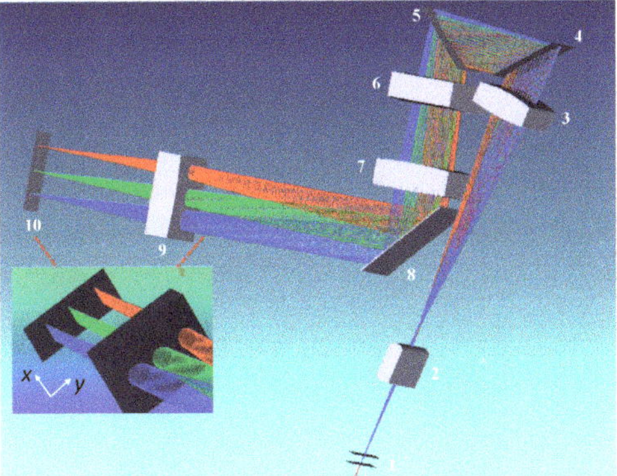

Figure 2. The proposed compact light path using dual-grating. The insets show the receiver beam distribution on the LCoS plane at a 45° view. The numbers 1–10 represent: No. 1 is the polarization converter with a birefringent crystal YVO_4 and half-wave plates; Nos. 2 and 3 are cylindrical mirrors with focal lengths of 4 mm and 60 mm, respectively; Nos. 4 and 5 are transmission gratings with 1000-line/mm resolution; Nos. 6, 7, and 9 are cylindrical lenses with focal length of 141 mm; No. 8 is a mirror; No. 10 is an LCoS chip, whose working wavelength is from 1520 nm to 1620 nm, pixel size is 8μm, fill factor is 87% and resolution is 1920 × 1080.

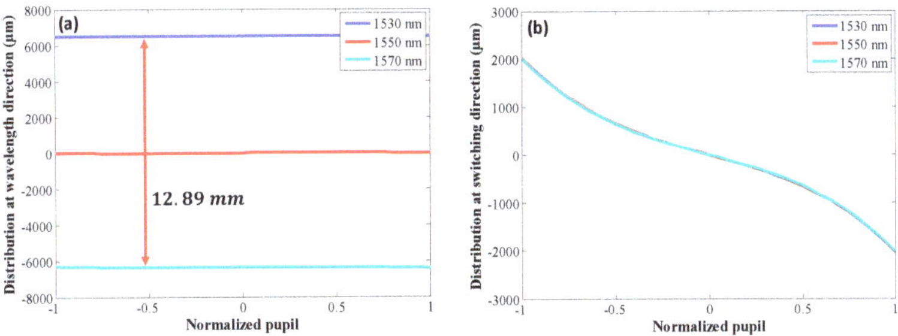

Figure 3. Simulation results using Zemax, (**a**) at the wavelength direction and (**b**) at the switching direction.

2.2.2. Experimental Setup

Figure 4 shows the experimental site of the assembled programmable PSP utilizing a compact light configuration, including a fiber array, a micro-lens array for the interaction of the light beam between the fiber and free space, a polarization conversion unit, collimating lenses, two transmission gratings, cemented lenses and a 2 K LCoS-SLM. The dense WDM optical signal from one fiber input port can be switched to several different output ports by the spectral dispersed gratings addressed on the LCoS. In the paper, one fiber port is picked as the input, and six ports with a 500 μm pitch are selected as outputs. The six output ports are divided into two groups, addressed equally on the upper and lower part of the input port. This design with symmetrical distribution helps to reduce IL differences between the output ports, because the closer the input port is to the center, the smaller the switching angle and the smaller the loss are. The beam divergence half-angle is compressed by the micro-lens array from around 0.6° to 6°. To maximally decrease the pointing error, the micro-lens array must be very carefully aligned with the input fiber array during the design process. Then, the input polarization states onto the LCoS chip are adjusted to the P-polarization, since the LCoS chip only manipulates the P-polarization light beams. A pair of cylindrical lenses are employed to collimate the beam spots and expand them into elliptical beams. The input beam further goes through the proposed light path by using dual gratings. The cemented lens, corresponding to a lens with a focal length of 100 mm, is composed of two cylindrical lenses ($f_1 = f_2 = 141$ mm). We add a mirror with a 45° reflective angle in the design to make the light path compact and more conveniently adjustable. The collimated and expanded light is finally converged on the LCoS display and modulated by its phase grating. Different optical spectra are projected onto different pixels of the LCoS chip plane. The diffracted and reflected spectra are recombined again by the transmission gratings onto the individual output ports. Thus, light beams with different wavelengths are able to be routed to any desired output ports.

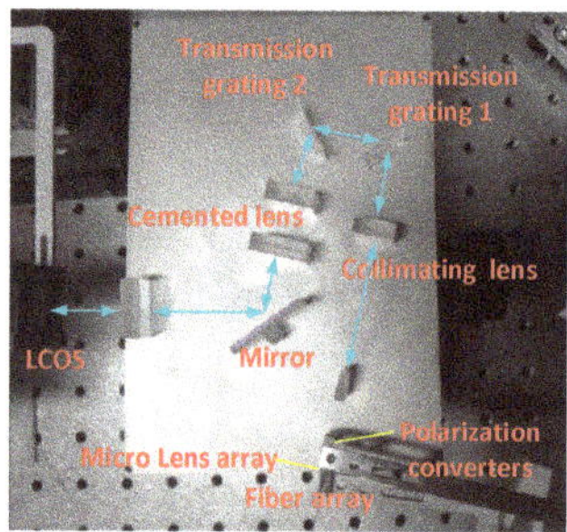

Figure 4. Experimental site using our proposed low-cost compact spatial light paths. All the optical elements are solidified on an optical substrate.

3. Experimental Results and Discussions

We first measured the IL of each output port as a function of the input wavelength across the whole C-band. In the experiment, a scanning laser source module (Agilent, 81940A) with a sweeping step of 0.001 nm is employed as a transmitter whose wavelength range is set from 1530 nm to 1570 nm. At the receiver side, we use a high-sensitivity optical power meter module (Agilent, N7745A) to detect the optical spectra one by one. According to the measured results as shown in Figure 5, the total IL fluctuations for the six output ports are between 5.9 dB to 9.4 dB, corresponding to port 1 and port 4. The introduced IL can be mainly classified as static loss and dynamic loss [24]. The former one is mainly caused by optical components such as the two transmission gratings (1.9 dB), micro-lens array (1.5 dB), LCoS chip (0.9 dB), collimating lens (0.5 dB), polarization conversion unit (0.5 dB), and cylindrical lens (0.2 dB). These ILs are calculated according to the corresponding material parameters. Due to the input and output of the optical path and the record of passing twice, these losses are doubled in the calculation. The latter one is mainly induced by different responses of different wavelengths passing through the slanted cascaded transmission gratings and different distances between the principal optical axis and output ports, leading to fluctuation losses from 0.4 dB to 3.9 dB. For each port, the measured intrinsic WDL within the full C-band is around 1 dB as illustrated in Figure 5.

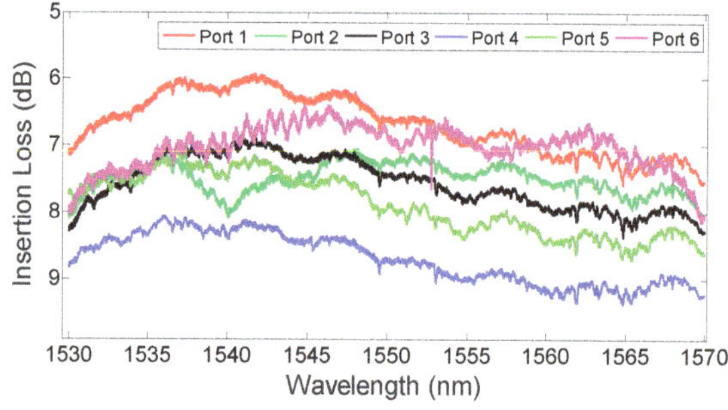

Figure 5. Measured total insertion loss (IL) for six ports in the full C-band wavelengths.

In order to facilitate the programming of the spectrum, we have written the corresponding operation software, and its operation interface is similar to that of Finisar WaveShaper. After calibrating the LCoS-SLM grayscale of the corresponding device port, and the accuracy of the wavelength and pixel control bandwidth, our software can also directly read the .txt or .xlsx data files composed of the wavelength, port and attenuation along the whole C-band, so as to realize the programming on the wavelength.

In the following measured results illustrated in Figures 6–9, we choose port 3 with neither the maximum nor the minimum IL of the C-band as the representative of the research object. Figure 6 shows the measured flexible 3-dB bandwidth adjustments from 6.2 GHz to 25.3 GHz by increasing the number of pixels on the LCoS chip one-by-one across the y-axis. The starting position is selected at the center of all the covered LCoS pixels, corresponding to the wavelength of 1550.8 nm. Without introducing extra IL, the achieved minimum 3-dB bandwidth is 9.8 GHz as shown in Figure 6. The maximum bandwidth can be extended to 5 THz across the entire C-band by adjusting a step of about 3.1 GHz, which is consistent with the theoretical minimum resolution of 3 GHz. The minimum filtering bandwidth of 6.2 GHz (corresponding to one pixel) can be obtained by introducing a 9-dB extra loss, since a fewer number of pixels results in a higher optical power loss. When increasing the 3-dB bandwidth to 8.1 GHz, the extra loss is decreased to 1.4 dB. We also measured the minimum 3-dB bandwidths in the C-band at the central wavelengths of 196 THz (1529.553 nm), 195 THz (1537.397 nm), 194 THz (1545.322 nm), 193 THz (1553.329 nm), 192 THz (1561.419 nm) and191 THz (1569.594 nm), respectively. As illustrated in Figure 7, the measured values are about 8.4 GHz, 8.1 GHz, 6.7 GHz, 6.59 GHz, 6.9 GHz and 7 GHz, corresponding to side band suppression ratios of 27 dB, 28 dB, 31 dB, 31.1 dB, 31.3 dB, and 29.3 dB, respectively.

Furthermore, a programmable PSP needs to possess the capability to arbitrarily attenuate the optical power in order to meet the special spectral requirement. In the experiment, by using our proposed PSP, we generate three sets of continuous sine waves with repetition frequencies of 2.5 GHz (0.02 nm), 0.25 GHz (0.002 nm), and 0.125 GHz (0.001 nm), respectively. As shown in Figure 8, the achieved peak-to-peak power suppressions for 2.5 GHz, 0.25 GHz and 0.125 GHz sine waves are 23.4 dB, 18 dB and 6 dB, respectively. The experimental results show that when the frequency interval is greater than or equal to 0.25 GHz, our proposed system can already produce a peak-to-peak power suppression signal greater than 18 dB, and the performance is good. In addition, in order to further verify the application of our method in communications such as OFDM, we have also produced OFCs with spacing of 12.5 GHz and 25 GHz.

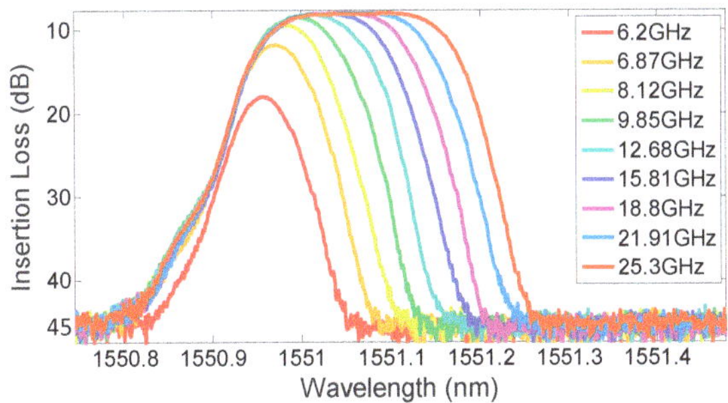

Figure 6. Flexible bandwidth adjustment with a minimum 3-dB bandwidth of 6.2 GHz to 25.3 GHz.

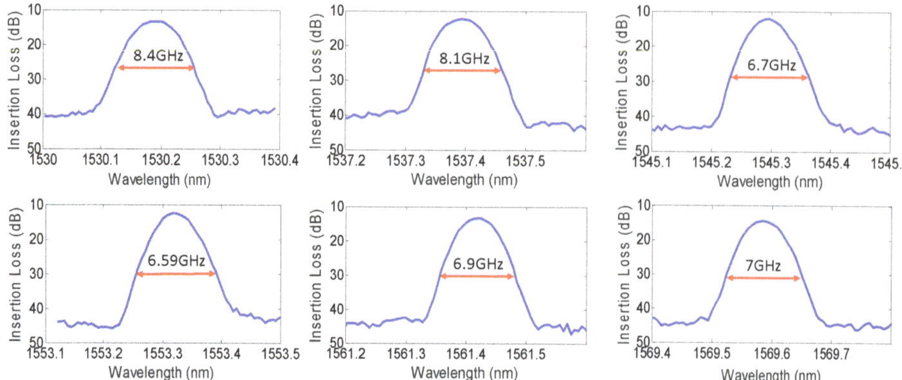

Figure 7. Measured minimum bandwidths in the C-band at the central wavelengths of 196 THz (1529.553 nm), 195 THz (1537.397 nm), 194 THz (1545.322 nm), 193THz (1553.329 nm), 192 THz (1561.419 nm) and 191 THz (1569.594 nm), respectively.

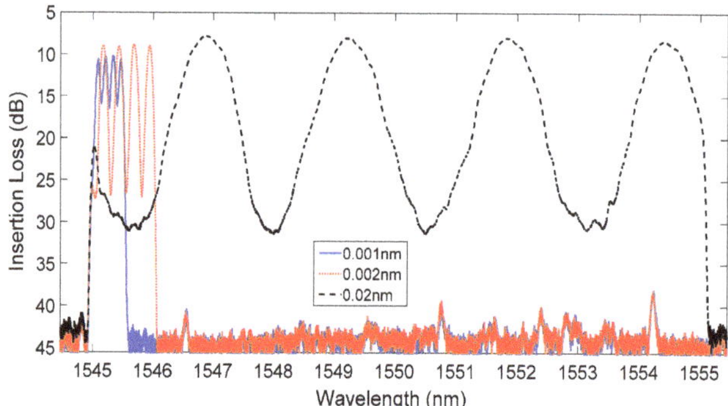

Figure 8. Measured optical spectra for 2.5 GHz (0.02 nm), 0.25 GHz (0.002 nm), and 0.125 GHz (0.001 nm) continuous sine waves.

Figure 9a shows the measured spectral results of 12.5 GHz and 25 GHz combs by using our designed 1 × 6 PSP, and the peak-to-peak power suppressions are more than 8 dB and 25 dB, respectively. Figure 9b shows the measured spectral results of 12.5 GHz and 25 GHz combs by using a commercial PSP, whose peak-to-peak power suppressions are about 3 dB and 18.7 dB, respectively. Obviously, the performance of our proposed system outperforms that of commercial PSP. This is due to the stronger ability by the proposed method to distinguish photons to the output ports.

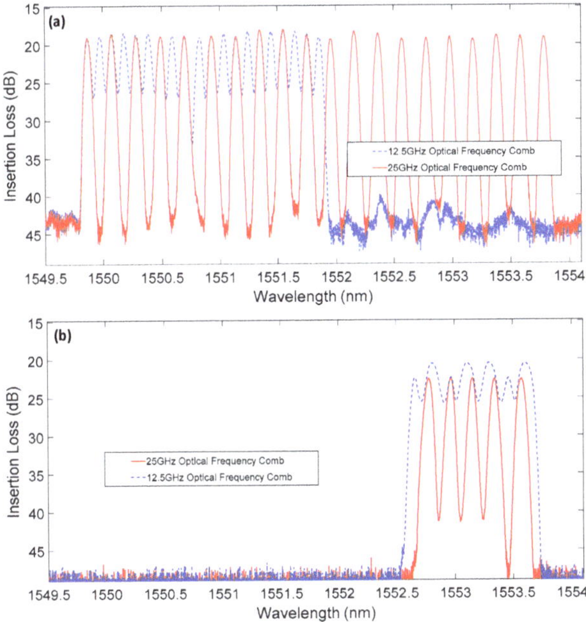

Figure 9. Measured spectra of optical frequency combs (OFCs) with 12.5 GHz and 25 GHz space using (**a**) our 1 × 6 PSP and (**b**) commercial PSP.

4. Conclusions

In this paper, we have developed a fine-resolution programmable PSP based on low-cost compact light paths, enabled by a 2K LCoS-SLM and dual cascaded 1000 line/mm diffraction gratings. An assembled experimental system with one-input and six-output has been built in order to implement a flexible filtering bandwidth and arbitrary amplitude transfer function. The measured total ILs for the six ports are around 5.9 dB–9.4 dB over the full C-band. The achieved minimum 3-dB bandwidth is 6.2 GHz, and the largest operating bandwidth can be extended to the full C-band from 1530 nm to 1570 nm (5 THz). Moreover, by using the proposed PSP, a high-resolution high-performance arbitrary optical spectrum can be generated, such as a sine wave signal with a 0.25-GHz repetition frequency and 18-dB peak-to-peak power suppression. Based on the achieved experimental results, developing a programmable and high-resolution 1 × N PSP by using low-cost compact spatial light paths is feasible, and is believed to further the implementation of an M × N PSP.

Author Contributions: Z.L. performed the investigation and experiment, wrote the original draft; C.L. performed the analytical calculations and revised the manuscript; J.T. discussed with the experimental results and revised the manuscript; S.Y. supervised the whole project and approved the manuscript. All authors have read and agreed to the published version of the manuscript.

Funding: This work was supported by National Natural Science Foundation of China (61805184), Hubei Provincial Natural Science Foundation of China (2018CFB250), Anhui Provincial Natural Science Foundation of China (1808085MF186).

Acknowledgments: The first author would like to thank Quan You, Ying Qiu and Xi Xiao from National Information Optoelectronics Innovation Center, Wuhan Research Institute of Posts and Telecommunications, for useful discussions during the experiment. And also thank Fulong Yan from Eindhoven University of Technology for valuable suggestions.

Conflicts of Interest: The authors declare no conflict of interest.

References

1. Zhang, J.; Yu, J.; Zhu, B.; Li, F.; Chien, H.; Jia, Z.; Cai, Y.; Li, X.; Xiao, X.; Fang, Y.; et al. Transmission of single-carrier 400G signals (515.2-Gb/s) based on 128.8-GBaud PDM QPSK over 10,130-and 6,078 km terrestrial fiber links. *Opt. Express* **2017**, *23*, 16540–16545. [CrossRef]
2. Marom, D.M.; Colbourne, P.D.; D'Errico, A.; Fontaine, N.K.; Ikuma, Y.; Proietti, R.; Zong, L.; Moscoso, J.; Tomkos, I. Survey of photonic switching architectures and technologies in support of spatially and spectrally flexible optical networking. *J. Opt. Commun. Netw.* **2017**, *9*, 1. [CrossRef]
3. Chandrasekhar, S.; Liu, X. OFDM based superchannel transmission technology. *IEEE/OSA J. Lightwave Technol.* **2012**, *30*, 3816–3823. [CrossRef]
4. Xiang, M.; Fu, S.; Tang, M.; Taang, H.; Shum, P.; Liu, D. Nyquist WDM superchannel using offset-16QAM and receiver-side digital spectral shaping. *Opt. Express* **2014**, *22*, 17448. [CrossRef] [PubMed]
5. Woodward, S.L.; Feuer, M.D. Benefits and requirements of flexible-grid ROADMs and networks. *J. Opt. Commun. Netw.* **2013**, *5*, A19. [CrossRef]
6. Li, Y.; Zong, L.; Gao, M.; Mukherjee, B.; Shen, G. Colorless, partially directional, and contentionless architecture for high-degree ROADMs. In Proceedings of the Optical Fiber Communications Conference and Exhibition (OFC), San Diego, CA, USA, 8–12 March 2020; IEEE: Piscataway, NJ, USA, 2020; p. M4D.2.
7. Yang, H.; Dolan, P.; Robertson, B.; Wilkinson, P.; Chu, D. Crosstalk spectrum optimisation for stacked wavelength selective switches based on 2D beam steering. In Proceedings of the Optical Fiber Communications Conference and Exhibition (OFC), San Diego, CA, USA, 11–15 March 2018; IEEE: Piscataway, NJ, USA, 2018; p. Th1J.2.
8. Wilkinson, P.; Robertson, B.; Giltrap, S.; Snowdon, O.; Prudden, H.; Yang, H.; Chu, D. 241 × 12 wavelength-selective switches using a 312-port 3D waveguide and a single 4k LCoS. In Proceedings of the Optical Fiber Communications Conference and Exhibition (OFC), San Diego, CA, USA, 8–12 March 2020; IEEE: Piscataway, NJ, USA, 2020; p. M3F.2.
9. Ma, Y.; Clarke, I.; Stewart, L. Recent progress on wavelength selective switch. In Proceedings of the Optical Fiber Communications Conference and Exhibition (OFC), San Diego, CA, USA, 8–12 March 2020; IEEE: Piscataway, NJ, USA, 2020; p. M3F.1.
10. Sorimoto, K.; Tanizawa, K.; Uetsuka, H.; Kawashima, H.; Mori, M.; Hasama, T.; Ishikawa, H.; Tsuda, H. Compact and phase-error-robust multilayered AWG-based wavelength selective switch driven by a single LCOS. *Opt. Express* **2013**, *24*, 17131–17149. [CrossRef] [PubMed]
11. Robertson, B.; Yang, H.; Redmond, M.; Collings, N.; Moore, J.R.; Liu, J.; Jeziorska-Chapman, A.M.; Pivnenko, M.; Lee, S.A.; Wonfor, I.H.; et al. Demonstration of multi-casting in a 1 × 9 lcos wavelength selective switch. *IEEE/OSA J. Lightwave Technol.* **2014**, *32*, 402. [CrossRef]
12. Iwama, M.; Takahashi, M.; Kimura, M.; Uchida, Y.; Hasegawa, J.; Hara, R.K.; Kagi, N. LCoS-based flexible grid 1 × 40 wavelength selective switch using planar lightwave circuit as spot size converter. In Proceedings of the Optical Fiber Communications Conference and Exhibition (OFC), Los Angeles, CA, USA, 22–26 March 2015; IEEE: Piscataway, NJ, USA, 2015; p. Tu3A.8.
13. Lu, T.; Collings, N.; Robertson, B.; Chu, D. Design of a low-cost and compact 1 × 5 wavelength-selective switch for access networks. *OSA Appl. Opt.* **2015**, *54*, 8844–8855. [CrossRef] [PubMed]
14. Xie, D.; Wang, D.; Zhang, M.; Liu, Z.; You, Q.; Yang, Q.; Yu, S. LCoS-based wavelength-selective switch for future finer-grid elastic optical networks capable of all-optical wavelength conversion. *IEEE Photonics J.* **2017**, *9*, 7101212. [CrossRef]
15. Gao, Y.; Chen, G.; Chen, X.; Zhang, Q.; Chen, Q.; Zhang, C.; Tian, K.; Tan, Z.; Yu, C. High-resolution tunable filter with flexible bandwidth and power attenuation based on an LCoS processor. *IEEE Photonics J.* **2018**, *10*, 6. [CrossRef]
16. Gao, Y.; Chen, X.; Chen, G.; Tan, Z.; Chen, Q.; Dai, D.; Zhang, Q.; Yu, C. Programmable spectral filter in C-Band based on digital micromirror device. *Micromachines* **2019**, *10*, 163. [CrossRef] [PubMed]
17. Rudnick, R.; Tolmachev, A.; Sinefeld, D.; Golani, O.; Ben-Ezra, S.; Nazarathy, M.; Marom, D.M. Sub-GHz resolution photonic spectral processor and its system applications. *IEEE/OSA J. Lightwave Technol.* **2017**, *35*, 2218. [CrossRef]
18. Xie, D.; Liu, Z.; You, Q.; Yu, S. Demonstration of a 3 × 4 tunable bandwidth WSS with tunable attenuation using compact spatial light paths. *Opt. Express* **2017**, *25*, 11173. [CrossRef] [PubMed]

19. Fontaine, N.K.; Ryf, R.; Neilson, D.T. N × M wavelength selective crossconnect with flexible passbands. In Proceedings of the Optical Fiber Communications Conference and Exhibition (OFC), Los Angeles, CA, USA, 4–8 March 2012; IEEE: Piscataway, NJ, USA, 2011; p. PDP5B.2.
20. Xiao, F.; Alameh, K. Opto-VLSI-based N × M wavelength selective switch. *OSA Opt. Express* **2013**, *21*, 18160. [CrossRef] [PubMed]
21. Lbsen Photonics, PING-Sample-083. Available online: https://ibsen.com/products/transmission-gratings/ping-telecom-%20gratings/telecom-c-band-gratings/ping-sample-083/ (accessed on 5 December 2020).
22. Palmer, C.A.; Loewen, E.G. *Diffraction Grating Handbook*; Thermo RGL: New York, NY, USA, 2002.
23. Born, M.; Wolf, E. Principles of optics: Electromagnetic theory of propagation, interference and diffraction of light. *Phys. Today* **2000**, *53*, 77. [CrossRef]
24. Wang, M.; Zong, L.; Mao, L.; Marquez, A.; Ye, Y.; Zhao, H.; Vaquero, F.J. LCoS SLM study and its application in wavelength selective switch. *Photonics* **2017**, *4*, 22. [CrossRef]

Publisher's Note: MDPI stays neutral with regard to jurisdictional claims in published maps and institutional affiliations.

MDPI
St. Alban-Anlage 66
4052 Basel
Switzerland
Tel. +41 61 683 77 34
Fax +41 61 302 89 18
www.mdpi.com

Photonics Editorial Office
E-mail: photonics@mdpi.com
www.mdpi.com/journal/photonics